Fundamentals of Crystallography

Fundamentals of Crystallography

Wesley Palmer

STATES
ACADEMIC PRESS
www.statesacademicpress.com

Fundamentals of Crystallography
Wesley Palmer
ISBN: 978-1-63989-226-6 (Hardback)

STATES
ACADEMIC PRESS

Published by States Academic Press,
109 South 5th Street,
Brooklyn, NY 11249, USA

Cataloging-in-Publication Data

Fundamentals of crystallography / Wesley Palmer.
p. cm.
Includes bibliographical references and index.
ISBN 978-1-63989-226-6
1. Crystallography. 2. Chemistry. I. Palmer, Wesley.
QD905.2 .F86 2022
548--dc23

Table of Contents

Preface

The purpose of this book is to help students understand the fundamental concepts of this discipline. It is designed to motivate students to learn and prosper. I am grateful for the support of my colleagues. I would also like to acknowledge the encouragement of my family.

A crystal, also known as crystalline solid is a solid material whose components like atoms, ions and molecules are arranged in highly ordered microscopic structure which forms a crystal lattice that is extendable in any direction. The field of science that deals with the determination of arrangement of atoms in crystalline solids is known as crystallography. It focuses on measuring the crystal structure. It also studies how crystals are formed. X-ray crystallography, and electron and neutron diffraction are the techniques used to determine the structure of the crystals. X-ray crystallography is the method of determining the molecular and atomic structures of the crystals. Electron diffraction is the technique of determining the wave nature of electrons. Neutron crystallography is used for determining the magnetic and atomic structure of the crystals. The topics included in this book related to the field of crystallography are of utmost significance and bound to provide incredible insights to readers. It is compiled in such a manner, that it will provide an in-depth knowledge about the theory and practice of crystallography. In this book, constant effort has been made to make the understanding of difficult concepts related to the field of crystallography as easy and informative as possible, for the readers.

A foreword for all the chapters is provided below:

Chapter - What is Crystallography?

Crystal, also known as crystalline solid, is a material whose constituents are arranged in a highly ordered microscopic structure. The branch of science that deals with the study of crystals, its formation, and bonding and arrangement of atoms in crystalline solids is known as crystallography. This chapter has been carefully written to provide an easy understanding of crystallography.

Chapter - Various Forms of Crystallography

There are different types of crystallography that are studied such as quantum crystallography, X-ray crystallography, cryo-bio crystallography, deep level transient crystallography, electron crystallography, nuclear magnetic resonance crystallography, etc. This chapter discusses the different types of crystallography in detail.

Chapter - Crystal Structure

Crystal structure is the ordered arrangement of atoms, molecules and ions in a crystal. It is used in materials science to characterize different materials and phase identification. All the aspects related to crystal structure have been carefully analyzed in this chapter.

Chapter - Crystal Systems

The space groups, lattices, point groups, crystal forms, etc. together form the crystal system. A few of its kinds are triclinic crystal system, monoclinic crystal system, tetragonal crystal system, hexagonal crystal system, cubic crystal system, orthorhombic crystal system, etc. This chapter delves into crystal system and its types for an in-depth understanding of the subject.

Chapter - Defects in Crystals

In crystallography, crystal defects are the discrepancies in the regular pattern, arrangement and organization of atoms of crystalline solids. Some of the defects are point defects, planar defects, intrinsic and extrinsic defects, Frenkel defects, interstitial defects, vacancy defects, etc. The topics elaborated in this chapter will help in gaining a better perspective about these crystal defects.

Chapter - Diverse Aspects of Crystallography

Diverse significant aspects are studied under the domain of crystallography which include crystallization. This chapter has been carefully written to provide an easy understanding of the important aspects of crystallography.

Wesley Palmer

1

What is Crystallography?

Crystal, also known as crystalline solid, is a material whose constituents are arranged in a highly ordered microscopic structure. The branch of science that deals with the study of crystals, its formation, and bonding and arrangement of atoms in crystalline solids is known as crystallography. This chapter has been carefully written to provide an easy understanding of crystallography.

Crystal

A crystal consists of matter that is formed from an ordered arrangement of atoms, molecules, or ions. The lattice that forms extends out in three dimensions. Because there are repeated units, crystals have recognizable structures. Large crystals display flat regions (faces) and well-defined angles.

Crystals with obvious flat faces are called euhedral crystals, while those lacking defined faces are called anhedral crystals. Crystals consisting of ordered arrays of atoms that aren't always periodic are called quasicrystals.

Examples of everyday materials you encounter as crystals are table salt (sodium chloride or halite crystals), sugar (sucrose), and snowflakes. Many gemstones are crystals, including quartz and diamond.

There are also many materials that resemble crystals but are actually polycrystals. Polycrystals form when microscopic crystals fuse together to form a solid. These materials do not consist of ordered lattices.

Examples of polycrystals include ice, many metal samples, and ceramics. Even less structure is displayed by amorphous solids, which have disordered internal structure. An example of an amorphous solid is glass, which may resemble a crystal when faceted, yet isn't one.

Chemical Bonds

The types of chemical bonds formed between atoms or groups of atoms in crystals depend on their size and electronegativity. There are four categories of crystals as grouped by their bonding:

- Covalent Crystals: Atoms in covalent crystals are linked by covalent bonds. Pure nonmetals form covalent crystals (e.g., diamond) as do covalent compounds (e.g., zinc sulfide).
- Molecular Crystals: Entire molecules are bonded to each other in an organized manner. A good example is a sugar crystal, which contains sucrose molecules.
- Metallic Crystals: Metals often form metallic crystals, where some of the valence electrons are free to move throughout the lattice. Iron, for example, can form different metallic crystals.
- Ionic Crystals: Electrostatic forces form ionic bonds. A classic example is a halite or salt crystal.

Crystal Lattices

There are seven systems of crystal structures, which are also called lattices or space lattices:

- Cubic or Isometric: This shape includes octahedrons and dodecahedrons as well as cubes.
- Tetragonal: These crystals form prisms and double pyramids. The structure is like a cubic crystal, except one axis is longer than the other.
- Orthorhombic: These are rhombic prisms and dipyramids that resemble tetragons but without square cross-sections.
- Hexagonal: Six-sided prisms with a hexagon cross section.
- Trigonal: These crystals have a three-fold axis.
- Triclinic: Triclinic crystals tend not to be symmetrical.
- Monoclinic: These crystals resemble skewed tetragonal shapes.

Lattices may have one lattice point per cell or more than one, yielding a total of 14 Bravais crystal lattice types. Bravais lattices, named for physicist and crystallographer Auguste Bravais, describe the three-dimensional array made by a set of discrete points.

A substance may form more than one crystal lattice. For example, water can form hexagonal ice (such as snowflakes), cubic ice, and rhombohedral ice. It can also form amorphous ice.

Carbon can form diamond (cubic lattice) and graphite (hexagonal lattice).

Formation of Crystals

The process of forming a crystal is called crystallization. Crystallization commonly occurs when a solid crystal grows from a liquid or solution.

As a hot solution cools or a saturated solution evaporates, particles draw close enough for chemical bonds to form. Crystals can also form from deposition directly from the gas phase. Liquid crystals possess particles oriented in an organized manner, like solid crystals, yet are able to flow.

Classification of Crystals

Although there is no unique way to classify all the crystalline solids found in nature, even then these are divided into four important types on the basis of chemical binding of the constituent atoms.

Ionic Crystals

These are formed by a combination of highly electropositive ions (cations) and highly electronegative ions (anions). Thus strong electrostatic force of attraction acts within the ionic crystals. Therefore, a large amount of energy is required to separate ions from one another. The type of the crystal lattice depends upon (i). The size of the ion (ii). The necessity for the preservation of electrical neutrality. Therefore alternate cations and anions in equivalent amount are arranged in the ionic crystal e.g. NaCl, KF, CsCl, $BaSO_4$ etc.

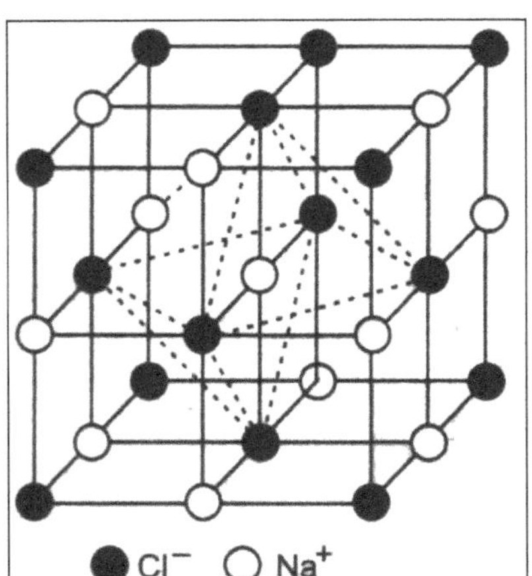

Covalent Crystals

These are formed by sharing of valence electrons between two atoms resulting in the formation of a covalent bond. The covalent bonds extend in two or three dimensions forming a giant interlocking structure called network. Diamond, silicon, quartz and graphite are good examples of this type.

Molecular Crystals

In these crystals, molecules occupy the lattice points of the unit cells, except in solidified noble gases in which the units are atoms, where the binding is due to van der Waals' forces and dipole-dipole forces. Since van der Waals' forces are non-directional, hence structure of the crystal is determined by geometric consideration only. Solids like H_2,O_2,CO_2,I_2,S_8, sugar etc. are well known examples of such crystals in which Van Der Waals forces are acting. Ice is the common example in which dipole-dipole forces of attraction (hydrogen bonding) are active. Many organic and inorganic crystals involve hydrogen bonds. Although these are comparatively weaker but they play a very important role in determining the structure of substances e. g. polynucleoides, proteins etc.

Metallic Crystals

These are formed by a combination of atoms of electropositive elements. These atoms are bound by metallic bonds. It may be defined as:

- The force that binds a metal ion to a number of electrons within its sphere of influence is known as metallic bond.

A bond which is formed between electropositive elements.

The attractive force which holds the atoms of two or more metals together in a metal crystal or in an alloy. We know that the force of attraction between metal ions and valency electrons is very strong. This force of attraction is responsible for a compact solid structure of metal.

Characteristics of Various Type of Crystals

Ionic Crystals

- Units that occupy lattice points: Cations and anions.
- Binding forces: Electrostatic attraction between ions.
- Hardness: Hard.
- Brittleness: Brittle.
- Melting point: High.
- Electrical Conduction: Semiconductor, due to crystal imperfections, conductor in fused state.
- Solubility in Polar solvents: Soluble.
- Heat of vaporization: NaCl(s).

  $kJmol^{-1}$: 170.75.

- Heat of Fusion: NaCl

  kJmol^{-1}: 28.45.

- Example: $NaCl, KNO_3, CsCl, Na_2SO_4, ZnS$.

Molecular Crystals

- Units that occupy lattice points: Molecules.
- Binding forces: Van der Waals or Dipole-dipole.
- Hardness: Soft.
- Brittleness: Low.
- Melting point: Low.
- Electrical Conduction: Bad conductor.
- Solubility in Polar solvents: Soluble as well as insoluble.
- Heat of vaporization: $NH_3(s)$.

  kJmol^{-1}: 23.55

- Heat of Fusion: $NH_3(s)$.

  kJmol^{-1}: 5.65

- Examples: $H_2O(s), CO_2(s)$, sulphur, sugar, iodine noble gases.

Metallic Crystals

- Units that occupy lattice points: positive ions in a "sea or pond" of electrons.
- Binding forces: Electrostatic attraction between positively charged ions and negatively charged electrons.
- Hardness: Hard or soft.
- Brittleness: Low.
- Melting point: Varying from moderate to high.
- Electrical Conduction: Good conductor.
- Solubility in Polar solvents: Insoluble.
- Heat of vaporization: Cu(s).

  kJmol^{-1}: 304.59.

- Heat of Fusion: Cu(s).

  $kJmol^{-1}$: 13.016.

- Example: Na, Cu, Ag, Fe, Pt, alloys.

Isomorphism

The occurrence of a given substance in more than one solid crystalline form having different physical properties is known as polymorphism. This property when occurs in elements is known as allotropy.

Sometimes we come across examples of chemically different solids which crystallise in the same crystalline shape. Such substances are said to be isomorphism (same shape), their chemical constitutions are very similar and in some cases crystals of one substance may continue to grow when placed in a saturated solution of the other e. g. potash alum and chrome alum crystals have the same shape and can be grown in each other's solutions.

Mitscherlich deduced that isomorphous substances have similar chemical formula e.g. phosphates and arsenates are said to be isomorphous with one another viz. $Na_2HPO_4.12H_2O$ and $Na_3AsO_4.12H_2O$. However, the law is not without exceptions.

Crystallography

Crystallography is the branch of science that deals with the geometric forms of crystals. How to describe, classify, and measure such forms are the first questions of crystallography. Revealing the forces that made them and the activities within them are the modern directions of the field. Crystallography is essential to progress in the applied sciences and technology and developments in all materials areas, including metals and alloys, ceramics, glasses, and polymers, as well as drug design. It is equally vital to progress in fundamental physics and chemistry, mineralogy and geology, and computer science, and to understanding of the dynamics and processes of living systems.

The external morphology of crystals reflects their growth rates in different directions. These directions remain constant during the course of the growth process, and are represented mathematically as the normals to sets of parallel planes that are imagined as being added on as growth proceeds. The faces that meet and define an edge belong to a zone, a zone being a set of planes that share one common direction, the zone axis. The invariance of interfacial angles, measured by rotation about an axis that is defined by the zone direction, was discovered in the seventeenth century.

Interfacial angles are calculated from spherical geometry. Figure below illustrates the procedure for a crystal having well-developed faces of which three are mutually perpendicular. The normals to these faces are the natural directions for constructing an orthogonal frame of reference for measurement. The crystal is imagined to be shrunk and placed at the center of a sphere with coordinates (0,0,0). The face normals, labeled,, and, define the directions of an orthogonal reference system. Normals to the same set of planes, but oppositely directed, are labeled [100], [1010], [001]. The reversal of sign indicates that the crystal must be rotated 180° to obtain the same view. Rotation about the direction interchanges the positions of and [100] faces and their bounding edges. Rotation about turns these faces upside down. Correct designations for group movements and symmetry operations are clearly essential for establishing and maintaining orientation in crystal space. The directions of face normals determine points at which the imagined sphere is pierced. The solid angles between an array of such points, all lying on the same great circle of the sphere, belong to a zone.

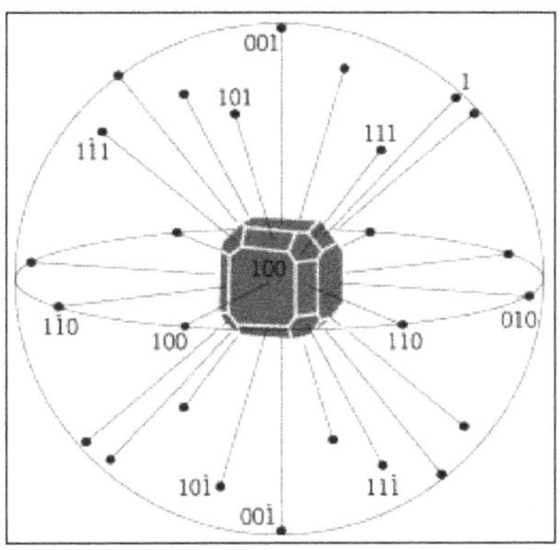

Spherical projection of normals to crystal faces.

Optical measurements and stereographic projections established the constancy of interfacial angles, independent of how well developed the faces are. Such properties as the cleavage of large rhombohedral crystals of calcite ($CaCO_3$) into little rhombs suggested that the large crystal could be represented by geometrically identical smaller units stacked together, by translation, to fill space. The 14 lattices of Bravais enlarged the seven crystal systems of optical mineralogy by adding centering points to them: body (I), face (F), and base (C) centers. The 14 lattices define three-dimensional distributions of mathematical points such that the environments of all points of the lattice are identical. They also define the symmetries of frameworks for constructing mathematical models to represent the observed and measured realities—models made from cells of the smallest volume, but also highest symmetry, that stack together by translation to fill space.

Table: The 14 Bravais lattices, derived by centering of the seven crystal classes (P and R) defined by symmetry operators.

Bravais lattice cells	Axes and interaxial angles	Example
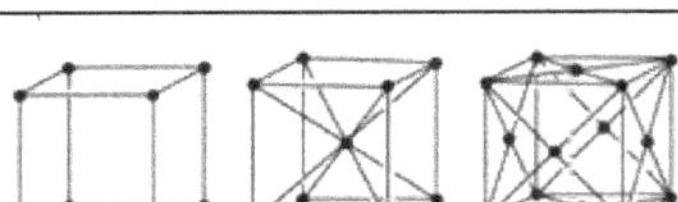 Cubic P Cubic I Cubic F	Three axes at right angles; all equal: a=b=c; $\alpha = \beta = \gamma = 90^\circ$	Copper (Cu), silver (Ag), sodium chloride (Nacl)

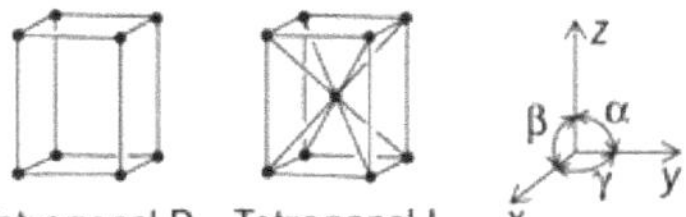 Tetragonal P Tetragonal I	Three axes at right angles; two equal: $a = b \neq c;\ \alpha = \beta = \gamma = 90^\circ$	White tin (Sn), rutile (TiO_2),β-spodumene ($LiAlSi_2O_6$)
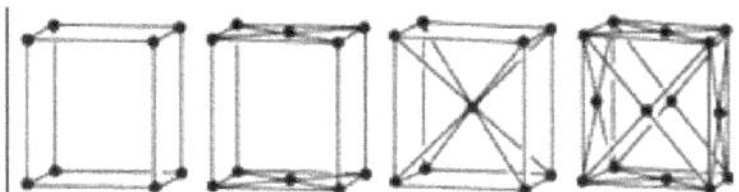 P C I F Orthorhombic	Three axes at right angles; all unequal: $a \neq b \neq c;\ \alpha = \beta = \gamma = 90^\circ$	Galium (Ga), pervoskite ($CaTiO_3$)
 Monoclinic P Monoclinic C	Three axes, one pair not at right angles, of any lengths: $a \neq b \neq c; \alpha = \gamma = 90^\circ \neq \beta$	Gypsum($CaSO_4.2H_2O$)
 Triclinic P	Three axes not at right angles, of any lengths: $a \neq b \neq c; \alpha \neq \beta \neq \gamma \neq 90^\circ$	Potassium chromate (K_2CrO_7)
 Trigonal R (rhombohedral)	Rhombohedral: Three axes equally inclined, not at right angles; all equal: $a = b = c; \alpha = \gamma = 90^\circ \neq \beta$	Calcite ($CaCO_3$),arsenic (As), bismuth (Bi)
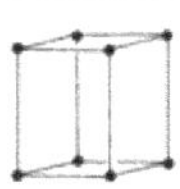 Trigonal and hexagonal C (or P)	Hexagonal: three equal axes coplanar 120°, fourth axis at right angles to these: $a_1, = a_2 = a_3 \neq c;$ $\alpha = \beta = 90^\circ, \gamma = 120^\circ$	Zinc (Zn), cadmium (Cd), quartz (SiO_2) [P]

Stacking of model cells does not imply that a crystal grows by stacking identical bricks; a lattice of identically surrounded mathematical points does not imply that any real objects, atoms or molecules, are located at the points; and filling space by translation of identical cells does not imply that the space defined by the cells is filled. Rather, the Bravais lattices are a formalism for representing observed geometries and symmetries of real crystals by three-dimensional lattices of identically surrounded points.

The lattices also provide the means to identify imaginary planes within the cell; these are called Miller indices (*h,k,l*). They consist of small whole numbers. For example, each of the six faces of a simple cube, with the origin of a coordinate frame of reference at the cube body center, is normal to one of the reference axes and parallel to the plane defined by the two others. The six faces are indexed as their normals in figure above—(100), (_00), (010), (0_0), (001), and (00_)—to represent a face that intercept the *x*, *x* axis but not the *y* and *z*; the *y*, *y* axis but not the *x* and *z*; and so forth. Hypothetical parallel planes with 1/2 the interplanar spacing are represented as (200), (_200), (020), and so forth.

A complete mathematical formalism exists for modeling an external morphological form and the symmetry relations between imagined units of structure within it. The symmetry operators include rotation axes, glide and mirror planes, and left- and right-handed screw axes which will simultaneously rotate and translate a three-dimensional object to create its clone in a different spatial position and orientation. The operators minimize the detail required to specify the spatial arrangements of patterns and objects that fill two-dimensional and three-dimensional space. The so-called color space groups of crystallography greatly increase the number of distinguishably different symmetries beyond the classical 230 by adding a fourth coordinate to the three space coordinates. This is done to encode a real difference that will be manifested in some property. The different directions of the magnetic moments of chemically identical atoms of an element such as iron provide an example of the need for representing a difference on the atomic level between cells that are otherwise identical.

The sharp x-ray line spectra characteristic of the bombarded element are the primary probes for determining interior structural detail of crystals. Cameras with cylindrical film and enclosed powdered samples record all diffraction lines as arcs of concentric circles. This fundamental powder method has endured since 1917 and is now employed with improved beam purity and optics, improved diffractometers which couple sample and detector rotation, electronic detection, rapid sequential recording, and computer indexing programs that provide patterns of compounds, mixtures of phases, and dynamic changes that occur when crystals are subjected and respond to external stress. The method is applied to single crystals, polycrystalline aggregates, and multiple-phase mixtures, randomly disposed either in space or in geometrically designed composite materials. The dynamics of living systems, the difficulties in distinguishing light elements, and the inherent ambiguities of measuring, decoding, and mapping crystal structures are continuing challenges. Major achievements of crystallography include the determination of the structures of deoxyribonucleic acid (DNA), proteins, other biological compounds, and boranes; the development of direct methods of phase determination; and the determination of the structure and mechanism of a photosynthetic center.

2
Various Forms of Crystallography

There are different types of crystallography that are studied such as quantum crystallography, X-ray crystallography, cryo-bio crystallography, deep level transient crystallography, electron crystallography, nuclear magnetic resonance crystallography, etc. This chapter discusses the different types of crystallography in detail.

Quantum Crystallography

Quantum crystallography is a branch of crystallography that investigates crystalline materials within the framework of quantum mechanics, with analysis and representation, in position or in momentum space, of quantities like wave function, electron charge and spin density, density matrices and all properties related to them (like electric potential, electric or magnetic moments, energy densities, electron localization function, one electron potential, etc.

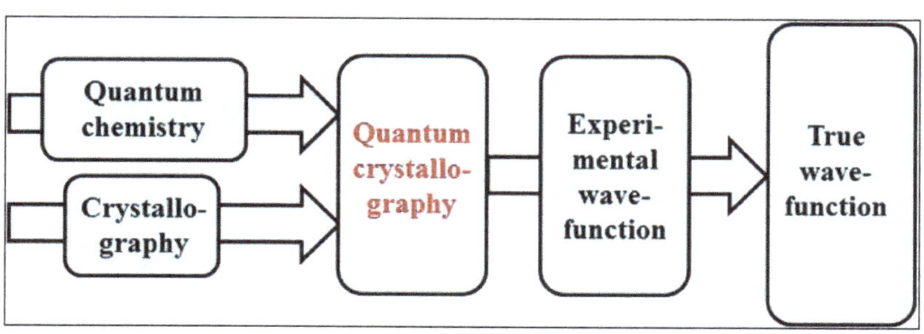

Quantum crystallography, first aspect: crystallographic data are integrated into quantum chemical calculations to enhance the information content of the wavefunction. The resulting, so-called "experimental wavefunction" represents an improved approximation to the true wavefunction.

The term quantum crystallography was first introduced by Massa, Huang and Karle in 1995 for methods that exploit "crystallographic information to enhance quantum mechanical calculations and the information derived from them". The basic idea is to compensate shortcomings and limitations of quantum mechanical models, e.g. incomplete consideration of electron correlation, with experimental data that are not suffering from the same limitations.

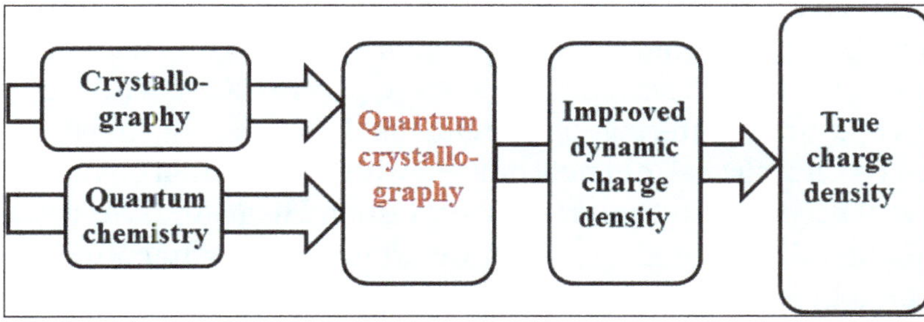

Quantum crystallography, second aspect: quantum chemical calculations are integrated into crystal structure determination to improve the dynamic charge density, i.e. the thermally smeared electron and nuclear densities.

Mutually subsidiary combinations of quantum chemistry and X-ray structure determination suggest themselves. Quantum chemical models are usually based on some approximations of the true wavefunction. X-ray structure determination aims at the true charge density, which is related to the square of the true wavefunction, but is affected to a smaller or larger extent by vibrational motion and experimental errors.

Pioneering "Experimental" Wavefunction Techniques

All modern quantum crystallography techniques according to the first definition originate from the pioneering "experimental" wavefunction strategies originally proposed in the 1960s by Mukherji and Karplus. They perturbed unconstrained Hartree–Fock molecular orbitals until they produced satisfactory agreement with experimental dipole moments or electric field gradients at a minimal increase in the energy of the system. Agreement improved not only for the constraining experimental values, but also for other properties, such as diamagnetic and paramagnetic susceptibilities. Rasiel and Whitman introduced experimental dipole moment constraints into the variational minimization of the energy with Lagrange multipliers. Byers Brown and Chong proposed to implicitly introduce the experimental constraints by adding proper quantum mechanical operators $\hat{O}_i$ (multiplied by a Lagrange multiplier λ_i) to the starting and unconstrained Hamiltonian $\hat{H}$:

$$\hat{H}' = \hat{H} + \sum_i \lambda_i \hat{O}_i$$

The modified Schrödinger equation can be solved variationally as shown by Fraga and Birss.

In 1969 Clinton et al. published a series of groundbreaking ideas that may be considered the direct precursors of today's quantum crystallography methods. Although X-ray diffraction data were not considered explicitly, the proposed theoretical framework forms the basis for the methods developed in the 1970s and 1980s that combine quantum mechanical calculations with crystallographic data. The researchers presented a semi-empirical strategy to determine one-electron density matrices of small diatomic molecules. They had both theoretical and practical reasons for focusing on one-electron density matrices rather than the more usual wavefunctions. To accomplish their task, they used theoretical constraints, such as the Hellmann–Feynman and virial theorems in combination with experimental constraints, e.g. potential energy curves obtained from spectroscopic and scattering measurements. However, the density matrices initially obtained by Clinton et al. did not conform with the essential quantum mechanical requirement that each density matrix must correspond to an antisymmetric wavefunction (N-representability). For single Slater determinant wavefunctions, this drawback can be overcome by imposing the idempotency condition (i.e., $\mathbf{P}^2 = \mathbf{P}$) and the normalization to the number of electrons (Tr$\mathbf{P}$ = N) on the one-electron density matrices. This goal was achieved by combining the original, iterative McWeeny density matrix purification formula:

$$\mathbf{P}_{n=1} = 3\mathbf{P}_n^2 - 2\mathbf{P}_n^3$$

with (i) the Hellmann–Feynman and virial theorems, (ii) Parr's integrated Hellmann–Feynman theorem, (iii) local energy constraints and, later, (iv) cusp conditions. This gave rise to another iterative procedure in which the density matrix at each iteration is obtained according to the following equation:

$$\mathbf{P}_{n+1} = 3\mathbf{P}_n^2 - 2\mathbf{P}_n^3 + \sum_k \lambda_k{}^{(n)}\mathbf{O}_k$$

where $\mathbf{P}_{n+1}$ and P n represent the one-electron density matrix in the adopted finite set of basis functions at the n+1 and n iterations; $\mathbf{O}_k$ is the matrix representation of the operator $\hat{O}_k$ associated with the property k in the same finite basis set as the density matrix. The external multipliers $\{\lambda_k{}^{(n)}\}$ are determined in a separate calculation by solving an auxiliary set of linear inhomogeneous equations that constrain the density matrix to reproduce the values {o k} of the properties considered.

X-ray Crystallography

X-ray crystallography (XRC) is the experimental science determining the atomic and molecular structure of a crystal, in which the crystalline structure causes a beam of incident X-rays to diffract into many specific directions. By measuring the angles and intensities of these diffracted beams, a crystallographer can produce a three-dimensional picture of the density of electrons within the crystal. From this electron density, the mean positions of the atoms in the crystal can be determined, as well as their chemical bonds, their crystallographic disorder, and various other information.

Since many materials can form crystals—such as salts, metals, minerals, semiconductors, as well as various inorganic, organic, and biological molecules—X-ray crystallography has been fundamental in the development of many scientific fields. In its first decades of use, this method determined the size of atoms, the lengths and types of chemical bonds, and the atomic-scale differences among various materials, especially minerals and alloys. The method also revealed the structure and function of many biological molecules, including vitamins, drugs, proteins and nucleic acids such as DNA. X-ray crystallography is still the primary method for characterizing the atomic structure of new materials and in discerning materials that appear similar by other experiments. X-ray crystal structures can also account for unusual electronic or elastic properties of a material, shed light on chemical interactions and processes, or serve as the basis for designing pharmaceuticals against diseases.

In a single-crystal X-ray diffraction measurement, a crystal is mounted on a goniometer. The goniometer is used to position the crystal at selected orientations. The crystal is illuminated with a finely focused monochromatic beam of X-rays, producing a diffraction pattern of regularly spaced spots known as *reflections*. The two-dimensional images taken at different orientations are converted into a three-dimensional model of the density of electrons within the crystal using the mathematical method of Fourier transforms, combined with chemical data known for the sample. Poor resolution (fuzziness) or even errors may result if the crystals are too small, or not uniform enough in their internal makeup.

X-ray crystallography is related to several other methods for determining atomic structures. Similar diffraction patterns can be produced by scattering electrons or neutrons, which are likewise interpreted by Fourier transformation. If single crystals of sufficient size cannot be obtained, various other X-ray methods can be applied to obtain less detailed information; such methods include fiber diffraction, powder diffraction and (if the sample is not crystallized) small-angle X-ray scattering (SAXS). If the material under investigation is only available in the form of nanocrystalline powders or suffers from poor crystallinity, the methods of electron crystallography can be applied for determining the atomic structure.

For all above mentioned X-ray diffraction methods, the scattering is elastic; the scattered X-rays have the same wavelength as the incoming X-ray. By contrast, *inelastic* X-ray scattering methods are useful in studying excitations of the sample such as plasmons, crystal-field and orbital excitations, magnons, and phonons, rather than the distribution of its atoms.

Contributions to Chemistry and Material Science

X-ray crystallography has led to a better understanding of chemical bonds and non-covalent interactions. The initial studies revealed the typical radii of atoms, and confirmed many theoretical models of chemical bonding, such as the tetrahedral bonding of carbon in the diamond structure, the octahedral bonding of metals observed in ammonium hexachloroplatinate (IV), and the resonance observed in the planar carbonate group and in aromatic molecules. Kathleen Lonsdale's 1928 structure of hexamethylbenzene established the hexagonal symmetry of benzene and showed a clear difference in bond length between the aliphatic C–C bonds and aromatic C–C bonds; this finding led to the idea of resonance between chemical bonds, which had profound consequences for the development of chemistry. Her conclusions were anticipated by William Henry Bragg, who published models of naphthalene and anthracene in 1921 based on other molecules, an early form of molecular replacement.

Also in the 1920s, Victor Moritz Goldschmidt and later Linus Pauling developed rules for eliminating chemically unlikely structures and for determining the relative sizes of atoms. These rules led to the structure of brookite and an understanding of the relative stability of the rutile, brookite and anatase forms of titanium dioxide.

The distance between two bonded atoms is a sensitive measure of the bond strength and its bond order; thus, X-ray crystallographic studies have led to the discovery of even more exotic types of bonding in inorganic chemistry, such as metal-metal double bonds, metal-metal quadruple bonds, and three-center, two-electron bonds. X-ray crystallography—or, strictly speaking, an inelastic Compton scattering experiment—has also provided evidence for the partly covalent character of hydrogen bonds. In the field of organometallic chemistry, the X-ray structure of ferrocene initiated scientific studies of sandwich compounds, while that of Zeise's salt stimulated research into "back bonding" and metal-pi complexes. Finally, X-ray crystallography had a pioneering role in the development of supramolecular chemistry, particularly in clarifying the structures of the crown ethers and the principles of host–guest chemistry.

X-ray diffraction is a very powerful tool in catalyst development. Ex-situ measurements are carried out routinely for checking the crystal structure of materials or to unravel new structures. In-situ experiments give comprehensive understanding about the structural stability of catalysts under reaction conditions.

In material sciences, many complicated inorganic and organometallic systems have been analyzed using single-crystal methods, such as fullerenes, metalloporphyrins, and other complicated compounds. Single-crystal diffraction is also used in the pharmaceutical industry, due to recent problems with polymorphs. The major factors affecting the quality of single-crystal structures are the crystal's size and regularity; recrystallization is a commonly used technique to improve these factors in small-molecule crystals. The Cambridge Structural Database contains over 1,000,000 structures as of June 2019; over 99% of these structures were determined by X-ray diffraction.

Mineralogy and Metallurgy

Since the 1920s, X-ray diffraction has been the principal method for determining the arrangement of atoms in minerals and metals. The application of X-ray crystallography to mineralogy began with the structure of garnet, which was determined in 1924 by Menzer. A systematic X-ray crystallographic study of the silicates was undertaken in the 1920s. This study showed that, as the Si/O ratio is altered, the silicate crystals exhibit significant changes in their atomic arrangements. Machatschki extended these insights to minerals in which aluminium substitutes for the silicon atoms of the silicates. The first application of X-ray crystallography to metallurgy likewise occurred in the mid-1920s. Most notably, Linus Pauling's structure of the alloy Mg_2Sn led to his theory of the stability and structure of complex ionic crystals.

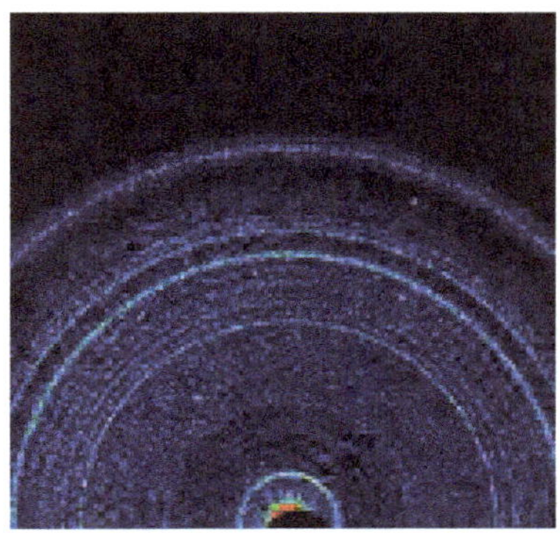

First X-ray diffraction view of Martian soil – CheMin analysis reveals feldspar, pyroxenes, olivine and more.

On October 17, 2012, the Curiosity rover on the planet Mars at "Rocknest" performed the first X-ray diffraction analysis of Martian soil. The results from the rover's CheMin analyzer revealed the presence of several minerals, including feldspar, pyroxenes and olivine, and suggested that the Martian soil in the sample was similar to the "weathered basaltic soils" of Hawaiian volcanoes.

Early Organic and Small Biological Molecules

The first structure of an organic compound, hexamethylenetetramine, was solved in 1923. This was followed by several studies of long-chain fatty acids, which are an important component of biological membranes. In the 1930s, the structures of much larger molecules with two-dimensional complexity began to be solved. A significant advance was the structure of phthalocyanine, a large planar molecule that is closely related to porphyrin molecules important in biology, such as heme, corrin and chlorophyll.

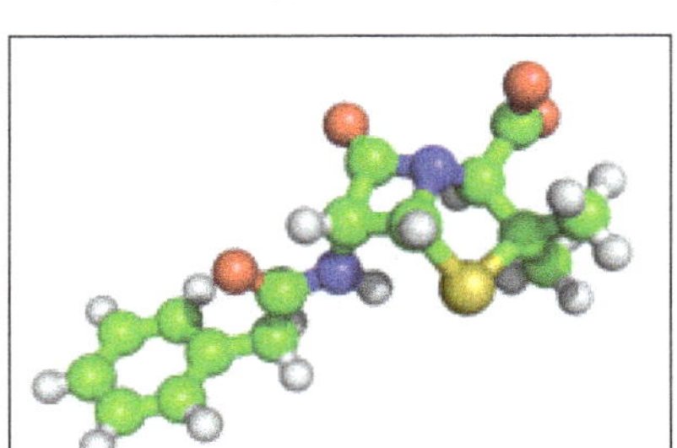

The three-dimensional structure of penicillin, solved by Dorothy Crowfoot Hodgkin in 1945. The green, red, yellow and blue spheres represent atoms of carbon, oxygen, sulfur and nitrogen, respectively. The white spheres represent hydrogen, which were determined mathematically rather than by the X-ray analysis.

X-ray crystallography of biological molecules took off with Dorothy Crowfoot Hodgkin, who solved the structures of cholesterol , penicillin and vitamin B_{12} , for which she was awarded the Nobel Prize in Chemistry in 1964. In 1969, she succeeded in solving the structure of insulin, on which she worked for over thirty years.

Biological Macromolecular Crystallography

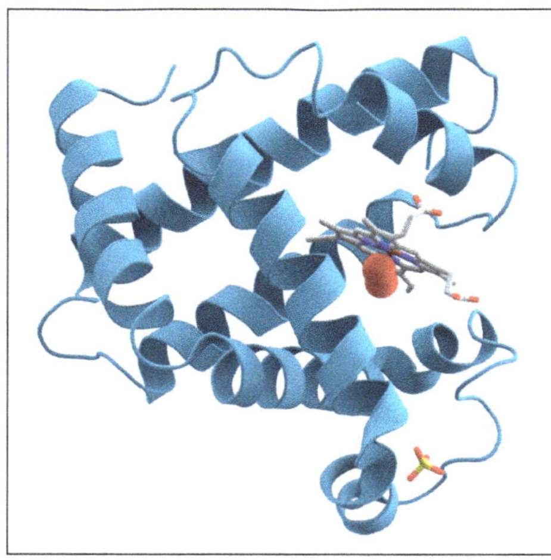

Ribbon diagram of the structure of myoglobin, showing colored alpha helices. Such proteins are long, linear molecules with thousands of atoms; yet the relative position of each atom has been determined with sub-atomic resolution by X-ray crystallography. Since it is difficult to visualize all the atoms at once, the ribbon shows the rough path of the protein polymer from its N-terminus (blue) to its C-terminus (red).

Crystal structures of proteins (which are irregular and hundreds of times larger than cholesterol) began to be solved in the late 1950s, beginning with the structure of sperm whale myoglobin by Sir John Cowdery Kendrew, for which he shared the Nobel Prize in Chemistry with Max Perutz in 1962. Since that success, over 130,000 X-ray crystal structures of proteins, nucleic acids and other biological molecules have been determined. The nearest competing method in number of structures analyzed is nuclear magnetic resonance (NMR) spectroscopy, which has resolved less than one tenth as many. Crystallography can solve structures of arbitrarily large molecules, whereas solution-state NMR is restricted to relatively small ones (less than 70 kDa). X-ray crystallography is used routinely to determine how a pharmaceutical drug interacts with its protein target and what changes might improve it. However, intrinsic membrane proteins remain challenging to crystallize because they require detergents or other denaturants to solubilize them in isolation, and such detergents often interfere with crystallization. Membrane proteins are a large component of the genome, and include many proteins of great physiological importance, such as ion channels and receptors. Helium cryogenics are used to prevent radiation damage in protein crystals.

On the other end of the size scale, even relatively small molecules may pose challenges for the resolving power of X-ray crystallography. The structure assigned in 1991 to the antibiotic isolated from a marine organism, diazonamide A ($C_{40}H_{34}Cl_2N_6O_6$, molar mass 765.65 g/mol), proved to be incorrect by the classical proof of structure: a synthetic sample was not identical to the natural product. The mistake was attributed to the inability of X-ray crystallography to distinguish between the correct -OH / -NH and the interchanged -NH_2 / -O- groups in the incorrect structure. With advances in instrumentation, however, similar groups can be distinguished using modern single-crystal X-ray diffractometers.

Despite being an invaluable tool in structural biology, protein crystallography carries some inherent problems in its methodology that hinder data interpretation. The crystal lattice, which is formed during the crystallization process, contains numerous units of the purified protein, which are densely and symmetrically packed in the crystal. When looking for a previously unknown protein, figuring out its shape and boundaries within the crystal lattice can be challenging. Proteins are usually composed of smaller subunits, and the task of distinguishing between the subunits and identifying the actual protein, can be challenging even for the experienced crystallographers. The non-biological interfaces that occur during crystallization are known as crystal-packing contacts (or simply, crystal contacts) and cannot be distinguished by crystallographic means. When a new protein structure is solved by X-ray crystallography and deposited in the Protein Data Bank, its authors are requested to specify the "biological assembly" which would constitute the functional, biologically-relevant protein. However, errors, missing data and inaccurate annotations during the submission of the data, give rise to obscure structures and compromise the reliability of the database. The error rate in the case of faulty annotations alone, has been reported to be upwards of 6,6% or approximately 15%, arguably a non-trivial size considering the number of deposited structures. This "interface classification problem" is typically tackled by computational approaches and has become a recognized subject in structural bioinformatics.

Scattering Techniques

Elastic vs. Inelastic Scattering

X-ray crystallography is a form of elastic scattering; the outgoing X-rays have the same energy, and thus same wavelength, as the incoming X-rays, only with altered direction. By contrast, inelastic scattering occurs when energy is transferred from the incoming

X-ray to the crystal, e.g., by exciting an inner-shell electron to a higher energy level. Such inelastic scattering reduces the energy (or increases the wavelength) of the outgoing beam. Inelastic scattering is useful for probing such excitations of matter, but not in determining the distribution of scatterers within the matter, which is the goal of X-ray crystallography.

X-rays range in wavelength from 10 to 0.01 nanometers; a typical wavelength used for crystallography is 1 Å (0.1 nm), which is on the scale of covalent chemical bonds and the radius of a single atom. Longer-wavelength photons (such as ultraviolet radiation) would not have sufficient resolution to determine the atomic positions. At the other extreme, shorter-wavelength photons such as gamma rays are difficult to produce in large numbers, difficult to focus, and interact too strongly with matter, producing particle-antiparticle pairs. Therefore, X-rays are the "sweetspot" for wavelength when determining atomic-resolution structures from the scattering of electromagnetic radiation.

Other X-ray Techniques

Other forms of elastic X-ray scattering besides single-crystal diffraction include powder diffraction, Small-Angle X-ray Scattering (SAXS) and several types of X-ray fiber diffraction, which was used by Rosalind Franklin in determining the double-helix structure of DNA. In general, single-crystal X-ray diffraction offers more structural information than these other techniques; however, it requires a sufficiently large and regular crystal, which is not always available.

These scattering methods generally use monochromatic X-rays, which are restricted to a single wavelength with minor deviations. A broad spectrum of X-rays (that is, a blend of X-rays with different wavelengths) can also be used to carry out X-ray diffraction, a technique known as the Laue method. This is the method used in the original discovery of X-ray diffraction. Laue scattering provides much structural information with only a short exposure to the X-ray beam, and is therefore used in structural studies of very rapid events. However, it is not as well-suited as monochromatic scattering for determining the full atomic structure of a crystal and therefore works better with crystals with relatively simple atomic arrangements.

The Laue back reflection mode records X-rays scattered backwards from a broad spectrum source. This is useful if the sample is too thick for X-rays to transmit through it. The diffracting planes in the crystal are determined by knowing that the normal to the diffracting plane bisects the angle between the incident beam and the diffracted beam. A Greninger chart can be used to interpret the back reflection Laue photograph.

Electron and Neutron Diffraction

Other particles, such as electrons and neutrons, may be used to produce a diffraction pattern. Although electron, neutron, and X-ray scattering are based on different physical processes, the resulting diffraction patterns are analyzed using the same coherent diffraction imaging techniques.

The electron density within the crystal and the diffraction patterns are related by a simple mathematical method, the Fourier transform, which allows the density to be calculated relatively easily from the patterns. However, this works only if the scattering is *weak*, i.e., if the scattered beams are much less intense than the incoming beam. Weakly scattered beams pass through the remainder of the crystal without undergoing a second scattering event. Such re-scattered waves are called "secondary scattering" and hinder the analysis. Any sufficiently thick crystal will produce secondary scattering, but since X-rays interact relatively weakly with the electrons, this is generally not a significant concern. By contrast, electron beams may produce strong secondary scattering even for relatively thin crystals (>100 nm). Since this thickness corresponds to the diameter of many viruses, a promising direction is the electron diffraction of isolated macromolecular assemblies, such as viral capsids and molecular machines, which may be carried out with a cryo-electron microscope. Moreover, the strong interaction of electrons with matter (about 1000 times stronger than for X-rays) allows determination of the atomic structure of extremely small volumes. The field of applications for electron crystallography ranges from bio molecules like membrane proteins over organic thin films to the complex structures of (nanocrystalline) intermetallic compounds and zeolites.

Neutron diffraction is an excellent method for structure determination, although it has been difficult to obtain intense, monochromatic beams of neutrons in sufficient quantities. Traditionally, nuclear reactors have been used, although sources producing neutrons by spallation are becoming increasingly available. Being uncharged, neutrons scatter much more readily from the atomic nuclei rather than from the electrons. Therefore, neutron scattering is very useful for observing the positions of light atoms with few electrons, especially hydrogen, which is essentially invisible in the X-ray diffraction. Neutron scattering also has the remarkable property that the solvent can be made invisible by adjusting the ratio of normal water, H_2O, and heavy water, D_2O.

Single-crystal X-ray Diffraction

The oldest and most precise method of X-ray crystallography is single-crystal X-ray diffraction, in which a beam of X-rays strikes a single crystal, producing scattered beams. When they land on a piece of film or other detector, these beams make a diffraction pattern of spots; the strengths and angles of these beams are recorded as the crystal is gradually rotated. Each spot is called a reflection, since it corresponds to the reflection of the X-rays from one set of evenly spaced planes within the crystal. For single crystals of sufficient purity and regularity, X-ray diffraction data can determine the mean chemical bond lengths and angles to within a few thousandths of an angstrom and to within a few tenths of a degree, respectively. The atoms in a crystal are not static, but oscillate about their mean positions, usually by less than a few tenths of an angstrom. X-ray crystallography allows measuring the size of these oscillations.

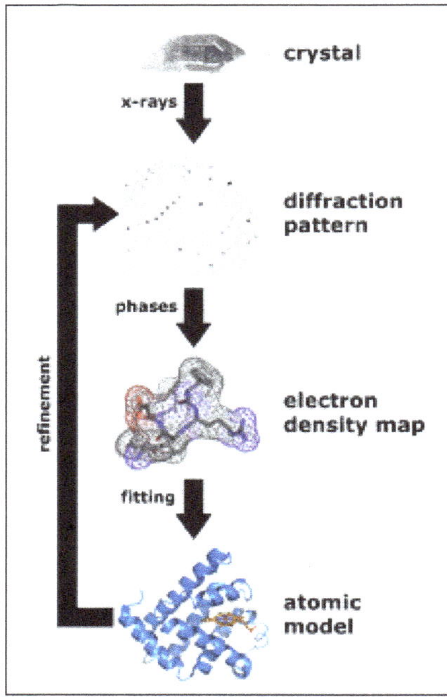

Workflow for solving the structure of a molecule by X-ray crystallography.

Procedure

The technique of single-crystal X-ray crystallography has three basic steps. The first—and often most difficult—step is to obtain an adequate crystal of the material under study. The crystal should be sufficiently large (typically larger than 0.1 mm in all dimensions), pure in composition and regular in structure, with no significant internal imperfections such as cracks or twinning.

In the second step, the crystal is placed in an intense beam of X-rays, usually of a single wavelength (*monochromatic X-rays*), producing the regular pattern of reflections. The angles and intensities of diffracted X-rays are measured, with each compound having a unique diffraction pattern. As the crystal is gradually rotated, previous reflections disappear and new ones appear; the intensity of every spot is recorded at every orientation of the crystal. Multiple data sets may have to be collected, with each set covering slightly more than half a full rotation of the crystal and typically containing tens of thousands of reflections.

In the third step, these data are combined computationally with complementary chemical information to produce and refine a model of the arrangement of atoms within the crystal. The final, refined model of the atomic arrangement—now called a *crystal structure*—is usually stored in a public database.

Limitations

As the crystal's repeating unit, its unit cell, becomes larger and more complex, the atomic-level picture provided by X-ray crystallography becomes less well-resolved (more "fuzzy") for a given number of observed reflections. Two limiting cases of X-ray crystallography—"small-molecule" (which includes continuous inorganic solids) and "macromolecular" crystallography—are often discerned. Small-molecule crystallography typically involves crystals with fewer than 100 atoms in their asymmetric unit; such crystal structures are usually so well resolved that the atoms can be discerned as isolated "blobs" of electron density. By contrast, macromolecular crystallography often involves tens of thousands of atoms in the unit cell. Such crystal structures are generally less well-resolved (more "smeared out"); the atoms and chemical bonds appear as tubes of electron density, rather than as isolated atoms. In general, small molecules are also easier to crystallize than macromolecules; however, X-ray crystallography has proven possible even for viruses and proteins with hundreds of thousands of atoms, through improved crystallographic imaging and technology. Though normally X-ray crystallography can only be performed if the sample is in crystal form, new research has been done into sampling non-crystalline forms of samples.

Crystallization

A protein crystal seen under a microscope. Crystals used in X-ray crystallography may be smaller than a millimeter across.

Although crystallography can be used to characterize the disorder in an impure or irregular crystal, crystallography generally requires a pure crystal of high regularity to solve the structure of a complicated arrangement of atoms. Pure, regular crystals can sometimes be obtained from natural or synthetic materials, such as samples of metals, minerals or other macroscopic materials. The regularity of such crystals can sometimes be improved with macromolecular crystal annealing and other methods. However, in many cases, obtaining a diffraction-quality crystal is the chief barrier to solving its atomic-resolution structure.

Small-molecule and macromolecular crystallography differ in the range of possible techniques used to produce diffraction-quality crystals. Small molecules generally have few degrees of conformational freedom, and may be crystallized by a wide range of methods, such as chemical vapor deposition and recrystallization. By contrast, macromolecules generally have many degrees of freedom and their crystallization must be carried out so as to maintain a stable structure. For example, proteins and larger RNA molecules cannot be crystallized if their tertiary structure has been unfolded; therefore, the range of crystallization conditions is restricted to solution conditions in which such molecules remain folded.

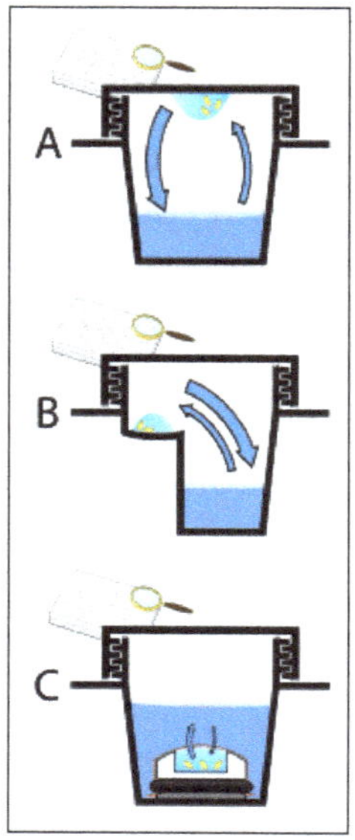

Three methods of preparing crystals, A: Hanging drop.
B: Sitting drop. C: Microdialysis.

Protein crystals are almost always grown in solution. The most common approach is to lower the solubility of its component molecules very gradually; if this is done too quickly, the molecules will precipitate from solution, forming a useless dust or amorphous gel on the bottom of the container. Crystal growth in solution is characterized by two steps: *nucleation* of a microscopic crystallite (possibly having only 100 molecules), followed by *growth* of that crystallite, ideally to a diffraction-quality crystal. The solution conditions that favor the first step (nucleation) are not always the same conditions that favor the second step (subsequent growth). The crystallographer's goal is to identify solution conditions that favor the development of a single, large crystal, since larger crystals offer improved resolution of the molecule. Consequently, the solution conditions should *disfavor* the first step (nucleation) but *favor* the second (growth), so that only one large crystal forms per droplet. If nucleation is favored too much, a shower of small crystallites will form in the droplet, rather than one large crystal; if favored too little, no crystal will form whatsoever. Other approaches involves, crystallizing proteins under oil, where aqueous protein solutions are dispensed under liquid oil, and water evaporates through the layer of oil. Different oils have different evaporation permeabilities, therefore yielding changes in concentration rates from different percipient/protein mixture.

It is extremely difficult to predict good conditions for nucleation or growth of well-ordered crystals. In practice, favorable conditions are identified by *screening*; a very large batch of the molecules is prepared, and a wide variety of crystallization solutions are tested. Hundreds, even thousands, of solution conditions are generally tried before finding the successful one. The various conditions can use one or more physical mechanisms to lower the solubility of the molecule; for example, some may change the pH, some contain salts of the Hofmeister series or chemicals that lower the dielectric constant of the solution, and still others contain large polymers such as polyethylene glycol that drive the molecule out of solution by entropic effects. It is also common to try several temperatures for encouraging crystallization, or to gradually lower the temperature so that the solution becomes supersaturated. These methods require large amounts of the target molecule, as they use high concentration of the molecule(s) to be crystallized. Due to the difficulty in obtaining such large quantities (milligrams) of crystallization-grade protein, robots have been developed that are capable of accurately dispensing crystallization trial drops that are in the order of 100 nanoliters in volume. This means that 10-fold less protein is used per experiment when compared to crystallization trials set up by hand (in the order of 1 microliter).

Several factors are known to inhibit or mar crystallization. The growing crystals are generally held at a constant temperature and protected from shocks or vibrations that might disturb their crystallization. Impurities in the molecules or in the crystallization solutions are often inimical to crystallization. Conformational flexibility in the molecule also tends to make crystallization less likely, due to entropy. Molecules that tend to self-assemble into regular helices are often unwilling to assemble into crystals. Crystals can be marred by twinning, which can occur when a unit cell can pack equally favorably in multiple orientations; although recent advances in computational methods may allow solving the structure of some twinned crystals. Having failed to crystallize a target molecule, a crystallographer may try again with a slightly modified version of the molecule; even small changes in molecular properties can lead to large differences in crystallization behavior.

Data Collection

Mounting the Crystal

Animation showing the five motions possible with a four-circle kappa goniometer. The rotations about each of the four angles φ, κ, ω and 2θ leave the crystal within the X-ray beam, but change the crystal orientation. The detector (red box) can be slid closer or further away from the crystal, allowing higher resolution data to be taken (if closer) or better discernment of the Bragg peaks (if further away).

The crystal is mounted for measurements so that it may be held in the X-ray beam and rotated. There are several methods of mounting. In the past, crystals were loaded into glass capillaries with the crystallization solution (the mother liquor). Nowadays, crystals of small molecules are typically attached with oil or glue to a glass fiber or a loop, which is made of nylon or plastic and attached to a solid rod. Protein crystals are scooped up by a loop, then flash-frozen with liquid nitrogen. This freezing reduces the radiation damage of the X-rays, as well as the noise in the Bragg peaks due to thermal motion (the Debye-Waller effect). However, untreated protein crystals often crack if flash-frozen; therefore, they are generally pre-soaked in a cryoprotectant solution before freezing. Unfortunately, this pre-soak may itself cause the crystal to crack, ruining it for crystallography. Generally, successful cryo-conditions are identified by trial and error.

The capillary or loop is mounted on a goniometer, which allows it to be positioned accurately within the X-ray beam and rotated. Since both the crystal and the beam are often very small, the crystal must be centered within the beam to within ~25 micrometers accuracy, which is aided by a camera focused on the crystal. The most common type of goniometer is the "kappa goniometer", which offers three angles of rotation: the ω angle, which rotates about an axis perpendicular to the beam; the κ angle, about an axis at ~50° to the ω axis; and, finally, the φ angle about the loop/capillary axis. When the κ angle is zero, the ω and φ axes are aligned. The κ rotation allows for convenient mounting of the crystal, since the arm in which the crystal is mounted may be swung out towards the crystallographer. The oscillations carried out during data collection (mentioned below) involve the ω axis only. An older type of goniometer is the four-circle goniometer, and its relatives such as the six-circle goniometer.

X-ray Sources

Rotating Anode

Small scale crystallography can be done with a local X-ray tube source, typically coupled with an image plate detector. These have the advantage of being relatively inexpensive and easy to maintain, and allow for quick screening and collection of samples. However, the wavelength of the light produced is limited by the availability of different anode materials. Furthermore, the intensity is limited by the power applied and cooling capacity available to avoid melting the anode. In such systems, electrons are boiled off of a cathode and accelerated through a strong electric potential of ~50 kV; having reached a high speed, the electrons collide with a metal plate, emitting *bremsstrahlung* and some strong spectral lines corresponding to the excitation of inner-shell electrons of the metal. The most common metal used is copper, which can be kept cool easily, due to its high thermal conductivity, and which produces strong K_α and K_β lines. The K_β line is sometimes suppressed with a thin (~10 μm) nickel foil. The simplest and cheapest variety of sealed X-ray tube has a stationary anode (the Crookes tube) and run with ~2 kW of electron beam power. The more expensive variety has a rotating-anode type source that run with ~14 kW of e-beam power.

X-rays are generally filtered (by use of X-ray filters) to a single wavelength (made monochromatic) and collimated to a single direction before they are allowed to strike the crystal. The filtering not only simplifies the data analysis, but also removes radiation that degrades the crystal without contributing useful information. Collimation is done either with a collimator (basically, a long tube) or with a clever arrangement of gently curved mirrors. Mirror systems are preferred for small crystals (under 0.3 mm) or with large unit cells (over 150 Å).

Rotating anodes were used by Joanna (Joka) Maria Vandenberg in the first experiments that demonstrated the power of X rays for quick (in real time production) screening of large InGaAsP thin film wafers for quality control of quantum well lasers.

Synchrotron Radiation

Synchrotron radiation sources are some of the brightest light sources on earth and are some of the most powerful tools available to X-ray crystallographers. X-ray beams generated in large machines called synchrotrons which accelerate electrically charged particles, often electrons, to nearly the speed of light and confine them in a (roughly) circular loop using magnetic fields.

Synchrotrons are generally national facilities, each with several dedicated beamlines where data is collected without interruption. Synchrotrons were originally designed for use by high-energy physicists studying subatomic particles and cosmic phenomena. The largest component of each synchrotron is its electron storage ring. This ring is actually not a perfect circle, but a many-sided polygon. At each corner of the polygon, or sector, precisely aligned magnets bend the electron stream. As the electrons' path is bent, they emit bursts of energy in the form of X-rays.

Using synchrotron radiation frequently has specific requirements for X-ray crystallography. The intense ionizing radiation can cause radiation damage to samples, particularly macromolecular crystals. Cryo crystallography protects the sample from radiation damage, by freezing the crystal at liquid nitrogen temperatures (~100 K). However, synchrotron radiation frequently has the advantage of user-selectable wavelengths, allowing for anomalous scattering experiments which maximizes anomalous signal. This is critical in experiments such as SAD and MAD.

Free-electron Laser

Free-electron lasers have been developed for use in X-ray crystallography. These are the brightest X-ray sources currently available; with the X-rays coming in femtosecond bursts. The intensity of the source is such that atomic resolution diffraction patterns can be resolved for crystals otherwise too small for collection. However, the intense light source also destroys the sample, requiring multiple crystals to be shot. As each crystal is randomly oriented in the beam, hundreds of thousands of individual diffraction images must be collected in order to get a complete data set. This method, serial femtosecond crystallography, has been used in solving the structure of a number of protein crystal structures, sometimes noting differences with equivalent structures collected from synchrotron sources.

Recording the Reflections

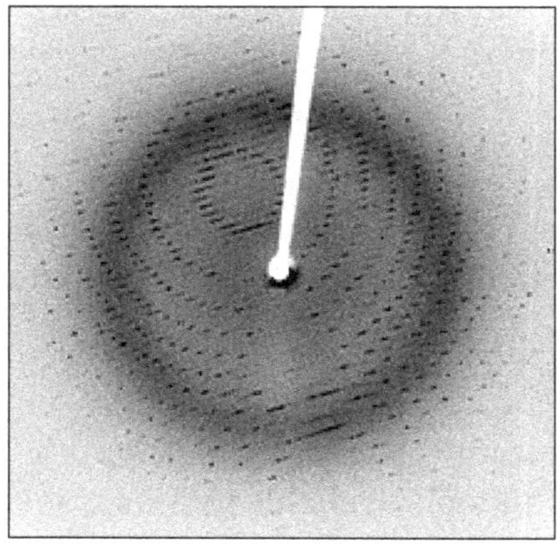

An X-ray diffraction pattern of a crystallized enzyme. The pattern of spots (*reflections*) and the relative strength of each spot (*intensities*) can be used to determine the structure of the enzyme.

When a crystal is mounted and exposed to an intense beam of X-rays, it scatters the X-rays into a pattern of spots or *reflections* that can be observed on a screen behind the crystal. A similar pattern may be seen by shining a laser pointer at a compact disc. The relative intensities of these spots provide the information to determine the arrangement of molecules within the crystal in atomic detail. The intensities of these reflections may be recorded with photographic film, an area detector (such as a pixel detector) or with a charge-coupled device (CCD) image sensor. The peaks at small angles correspond to low-resolution data, whereas those at high angles represent high-resolution data; thus, an upper limit on the eventual resolution of the structure can be determined from the first few images. Some measures of diffraction quality can be determined at this point, such as the mosaicity of the crystal and its overall disorder, as observed in the peak widths. Some pathologies of the crystal that would render it unfit for solving the structure can also be diagnosed quickly at this point.

One image of spots is insufficient to reconstruct the whole crystal; it represents only a small slice of the full Fourier transform. To collect all the necessary information, the crystal must be rotated step-by-step through 180°, with an image recorded at every step; actually, slightly more than 180° is required to cover reciprocal space, due to the curvature of the Ewald sphere. However, if the crystal has a higher symmetry, a smaller angular range such as 90° or 45° may be recorded. The rotation axis should be changed at least once, to avoid developing a "blind spot" in reciprocal space close to the rotation axis. It is customary to rock the crystal slightly (by 0.5–2°) to catch a broader region of reciprocal space.

Multiple data sets may be necessary for certain phasing methods. For example, MAD phasing requires that the scattering be recorded at least three (and usually four, for redundancy) wavelengths of the incoming X-ray radiation. A single crystal may degrade too much during the collection of one data set, owing to radiation damage; in such cases, data sets on multiple crystals must be taken.

Data Analysis

Crystal Symmetry, Unit Cell and Image Scaling

The recorded series of two-dimensional diffraction patterns, each corresponding to a different crystal orientation, is converted into a three-dimensional model of the electron density; the conversion uses the mathematical technique of Fourier transforms, which is explained below. Each spot corresponds to a different type of variation in the electron density; the crystallographer must determine *which* variation corresponds to *which* spot (*indexing*), the relative strengths of the spots in different images (*merging and scaling*) and how the variations should be combined to yield the total electron density (*phasing*).

Data processing begins with *indexing* the reflections. This means identifying the dimensions of the unit cell and which image peak corresponds to which position in reciprocal space. A byproduct of indexing is to determine the symmetry of the crystal, i.e., its *space group*. Some space groups can be eliminated from the beginning. For example, reflection symmetries cannot be observed in chiral molecules; thus, only 65 space groups of 230 possible are allowed for protein molecules which are almost always chiral. Indexing is generally accomplished using an *autoindexing* routine. Having assigned symmetry, the data is then *integrated*. This converts the hundreds of images containing the thousands of reflections into a single file, consisting of (at the very least) records of the Miller index of each reflection, and an intensity for each reflection (at this state the file often also includes error estimates and measures of partiality (what part of a given reflection was recorded on that image)).

A full data set may consist of hundreds of separate images taken at different orientations of the crystal. The first step is to merge and scale these various images, that is, to identify which peaks appear in two or more images (*merging*) and to scale the relative images so that they have a consistent intensity scale. Optimizing the intensity scale is critical because the relative intensity of the peaks is the key information from which the structure is determined. The repetitive technique of crystallographic data collection and the often high symmetry of crystalline materials cause the diffractometer to record many symmetry-equivalent reflections multiple times. This allows calculating the symmetry-related R-factor, a reliability index based upon how similar are the measured intensities of symmetry-equivalent reflections, thus assessing the quality of the data.

Initial Phasing

The data collected from a diffraction experiment is a reciprocal space representation of the crystal lattice. The position of each diffraction 'spot' is governed by the size and shape of the unit cell, and the inherent symmetry within the crystal. The intensity of each diffraction 'spot' is recorded, and this intensity is proportional to the square of the *structure factor* amplitude. The structure factor is a complex number containing information relating to both the amplitude and phase of a wave. In order to obtain an interpretable *electron density map*, both amplitude and phase must be known (an electron density map allows a crystallographer to build a starting model of the molecule). The phase cannot be directly recorded during a diffraction experiment: this is known as the phase problem. Initial phase estimates can be obtained in a variety of ways:

- Ab initio phasing or direct methods: This is usually the method of choice for small molecules (<1000 non-hydrogen atoms), and has been used successfully to solve the phase problems for small proteins. If the resolution of the data is better than 1.4 Å (140 pm), direct methods can be used to obtain phase information, by exploiting known phase relationships between certain groups of reflections.

- Molecular replacement: If a related structure is known, it can be used as a search model in molecular replacement to determine the orientation and position of the molecules within the unit cell. The phases obtained this way can be used to generate electron density maps.

- Anomalous X-ray scattering (MAD or SAD phasing): The X-ray wavelength may be scanned past an absorption edge of an atom, which changes the scattering in a known way. By recording full sets of reflections at three different wavelengths (far below, far above and in the middle of the absorption edge) one can solve for the substructure of the anomalously diffracting atoms and hence the structure of the whole molecule. The most popular method of incorporating anomalous scattering atoms into proteins is to express the protein in a methionine auxotroph (a host incapable of synthesizing methionine) in a media rich in seleno-methionine, which contains selenium atoms. A MAD experiment can then be conducted around the absorption edge, which should then yield the position of any methionine residues within the protein, providing initial phases.

- Heavy atom methods (multiple isomorphous replacement): If electron-dense metal atoms can be introduced into the crystal, direct methods or Patterson-space methods can be used to determine their location and to obtain initial phases. Such heavy atoms can be introduced either by soaking the crystal in a heavy atom-containing solution, or by co-crystallization (growing the crystals in the presence of a heavy atom). As in MAD phasing, the changes in the scattering amplitudes can be interpreted to yield the phases. Although this is the original method by which protein crystal structures were solved, it has largely been superseded by MAD phasing with selenomethionine.

Model Building and Phase Refinement

Having obtained initial phases, an initial model can be built. The atomic positions in the model and their respective Debye-Waller factors (or B-factors, accounting for the thermal motion of the atom) can be refined to fit the observed diffraction data, ideally yielding a better set of phases. A new model can then be fit to the new electron density map and successive rounds of refinement is carried out. This interative process continues until the correlation between the diffraction data and the model is maximized. The agreement is measured by an R-factor defined as:

$$R = \frac{\sum_{\text{all reflections}} \left| F_{\text{obs}} - F_{\text{calc}} \right|}{\sum_{\text{all reflections}} \left| F_{\text{obs}} \right|},$$

where F is the structure factor. A similar quality criterion is R_{free}, which is calculated from a subset (~10%) of reflections that were not included in the structure refinement. Both R factors depend on the resolution of the data. As a rule of thumb, R_{free} should be approximately the resolution in angstroms divided by 10; thus, a data-set with 2 Å resolution should yield a final $R_{\text{free}} \sim 0.2$. Chemical bonding features such as stereochemistry, hydrogen bonding and distribution of bond lengths and angles are complementary measures of the model quality. Phase bias is a serious problem in such iterative model building. *Omit maps* are a common technique used to check for this.

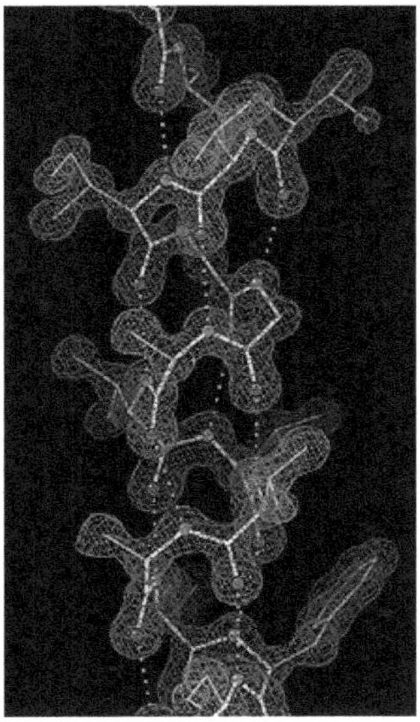

Structure of a protein alpha helix, with stick-figures for the covalent bonding within electron density for the crystal structure at ultra-high-resolution (0.91 Å). The density contours are in gray, the helix backbone in white, sidechains in cyan, O atoms in red, N atoms in blue, and hydrogen bonds as green dotted lines.

It may not be possible to observe every atom in the asymmetric unit. In many cases, Crystallographic disorder smears the electron density map. Weakly scattering atoms such as hydrogen are routinely invisible. It is also possible for a single atom to appear multiple times in an electron density map, e.g., if a protein sidechain has multiple (<4) allowed conformations. In still other cases, the crystallographer may detect that the covalent structure deduced for the molecule was incorrect, or changed. For example, proteins may be cleaved or undergo post-translational modifications that were not detected prior to the crystallization.

Disorder

A common challenge in refinement of crystal structures results from crystallographic disorder. Disorder can take many forms but in general involves the coexistence of two or more species or conformations. Failure to recognize disorder results in flawed interpretation. Pitfalls from improper modeling of disorder are illustrated by the discounted hypothesis of bond stretch isomerism. Disorder is modelled with respect to the relative population of the components, often only two, and their identity. In structures of large molecules and ions, solvent and counterions are often disordered.

Applied Computational Data Analysis

The use of computational methods for the powder X-ray diffraction data analysis is now generalized. It typically compares the experimental data to the simulated diffractogram of a model structure, taking into account the instrumental parameters, and refines the structural or microstructural parameters of the model using least squares based minimization algorithm. Most available tools allowing phase identification and structural refinement are based on the Rietveld method, some of them being open and free software such as FullProf Suite, Jana2006, MAUD, Rietan, GSAS, etc. while others are available under commercials licenses such as Diffrac.Suite TOPAS, Match!, etc. Most of these tools also allow Le Bail refinement (also referred to as profile matching), that is, refinement of the cell parameters based on the Bragg peaks positions and peak profiles, without taking into account the crystallographic structure by itself. More recent tools allow the refinement of both structural and microstructural data, such as the FAULTS program included in the FullProf Suite, which allows the refinement of structures with planar defects (e.g. stacking faults, twinnings, intergrowths).

Deposition of the Structure

Once the model of a molecule's structure has been finalized, it is often deposited in a crystallographic database such as the Cambridge Structural Database (for small molecules), the Inorganic Crystal Structure Database (ICSD) (for inorganic compounds) or the Protein Data Bank (for protein and sometimes nucleic acids). Many structures obtained in private commercial ventures to crystallize medicinally relevant proteins are not deposited in public crystallographic databases.

Diffraction Theory

The main goal of X-ray crystallography is to determine the density of electrons $f(\mathbf{r})$ throughout the crystal, where $\mathbf{r}$ represents the three-dimensional position vector within the crystal. To do this, X-ray scattering is used to collect data about its Fourier transform $F(\mathbf{q})$, which is inverted mathematically to obtain the density defined in real space, using the formula:

$$f(\mathbf{r}) = \frac{1}{(2\pi)^3} \int F(\mathbf{q}) e^{i\mathbf{q}\cdot\mathbf{r}} d\mathbf{q},$$

where the integral is taken over all values of $\mathbf{q}$. The three-dimensional real vector $\mathbf{q}$ represents a point in reciprocal space, that is, to a particular oscillation in the electron density as one moves in the direction in which $\mathbf{q}$ points. The length of $\mathbf{q}$ corresponds to non-target 2π divided by the wavelength of the oscillation. The corresponding formula for a Fourier transform will be used below:

$$F(\mathbf{q}) = \int f(\mathbf{r}) e^{-i\mathbf{q}\cdot\mathbf{r}} d\mathbf{r},$$

where the integral is summed over all possible values of the position vector $\mathbf{r}$ within the crystal.

The Fourier transform $F(\mathbf{q})$ is generally a complex number, and therefore has a magnitude $|F(\mathbf{q})|$ and a phase $\varphi(\mathbf{q})$ related by the equation:

$$F(\mathbf{q}) = |F(\mathbf{q})| e^{i\phi(\mathbf{q})}.$$

The intensities of the reflections observed in X-ray diffraction give us the magnitudes $|F(\mathbf{q})|$ but not the phases $\varphi(\mathbf{q})$. To obtain the phases, full sets of reflections are collected with known alterations to the scattering, either by modulating the wavelength past a certain absorption edge or by adding strongly scattering (i.e., electron-dense) metal atoms such as mercury. Combining the magnitudes and phases yields the full Fourier transform $F(\mathbf{q})$, which may be inverted to obtain the electron density $f(\mathbf{r})$.

Crystals are often idealized as being *perfectly* periodic. In that ideal case, the atoms are positioned on a perfect lattice, the electron density is perfectly periodic, and the Fourier transform $F(\mathbf{q})$ is zero except when $\mathbf{q}$ belongs to the reciprocal lattice (the so-called *Bragg peaks*). In reality, however, crystals are not perfectly periodic; atoms vibrate about their mean position, and there may be disorder of various types, such as mosaicity, dislocations, various point defects, and heterogeneity in the conformation of crystallized molecules. Therefore, the Bragg peaks have a finite width and there may be significant *diffuse scattering*, a continuum of scattered X-rays that fall between the Bragg peaks.

Intuitive Understanding by Bragg's Law

An intuitive understanding of X-ray diffraction can be obtained from the Bragg model of diffraction. In this model, a given reflection is associated with a set of evenly spaced sheets running through the crystal, usually passing through the centers of the atoms of the crystal lattice. The orientation of a particular set of sheets is identified by its three Miller indices (h, k, l), and let their spacing be noted by d. William Lawrence Bragg proposed a model in which the incoming X-rays are scattered specularly (mirror-like) from each plane; from that assumption, X-rays scattered from adjacent planes will combine constructively (constructive interference) when the angle θ between the plane and the X-ray results in a path-length difference that is an integer multiple n of the X-ray wavelength λ.

$$2d\sin\theta = n\lambda.$$

A reflection is said to be *indexed* when its Miller indices (or, more correctly, its reciprocal lattice vector components) have been identified from the known wavelength and the scattering angle 2θ. Such indexing gives the unit-cell parameters, the lengths and angles of the unit-cell, as well as its space group. Since Bragg's law does not interpret the relative intensities of the reflections, however, it is generally inadequate to solve for the arrangement of atoms within the unit-cell; for that, a Fourier transform method must be carried out.

Scattering as a Fourier Transform

The incoming X-ray beam has a polarization and should be represented as a vector wave; however, for simplicity, let it be represented here as a scalar wave. We also ignore the complication of the time dependence of the wave and just concentrate on the wave's spatial dependence. Plane waves can be represented by a wave vector $\mathbf{k}_{in}$, and so the strength of the incoming wave at time $t = 0$ is given by:

$$Ae^{i\mathbf{k}_{in}\cdot\mathbf{r}}.$$

At position $\mathbf{r}$ within the sample, let there be a density of scatterers $f(\mathbf{r})$; these scatterers should produce a scattered spherical wave of amplitude proportional to the local amplitude of the incoming wave times the number of scatterers in a small volume dV about $\mathbf{r}$.

$$\text{amplitude of scattered wave} = Ae^{i\mathbf{k}_{in}\cdot\mathbf{r}}Sf(\mathbf{r})dV,$$

where S is the proportionality constant.

Consider the fraction of scattered waves that leave with an outgoing wave-vector of $\mathbf{k}_{out}$ and strike the screen at $\mathbf{r}_{screen}$. Since no energy is lost (elastic, not inelastic scattering), the wavelengths are the same as are the magnitudes of the wave-vectors $|\mathbf{k}_{in}|=|\mathbf{k}_{out}|$. From the time that the photon is scattered at $\mathbf{r}$ until it is absorbed at $\mathbf{r}_{screen}$, the photon undergoes a change in phase,

$$e^{i\mathbf{k}_{out}\cdot(\mathbf{r}_{screen}-\mathbf{r})}.$$

The net radiation arriving at $\mathbf{r}_{\text{screen}}$ is the sum of all the scattered waves throughout the crystal,

$$AS\int d\mathbf{r} f(\mathbf{r}) e^{i\mathbf{k}_{\text{in}}\cdot\mathbf{r}} e^{i\mathbf{k}_{\text{out}}\cdot(\mathbf{r}_{\text{screen}}-\mathbf{r})} = ASe^{i\mathbf{k}_{\text{out}}\cdot\mathbf{r}_{\text{screen}}} \int d\mathbf{r} f(\mathbf{r}) e^{i(\mathbf{k}_{\text{in}}-\mathbf{k}_{\text{out}})\cdot\mathbf{r}},$$

which may be written as a Fourier transform,

$$ASe^{i\mathbf{k}_{\text{out}}\cdot\mathbf{r}_{\text{screen}}} \int d\mathbf{r} f(\mathbf{r}) e^{-i\mathbf{q}\cdot\mathbf{r}} = ASe^{i\mathbf{k}_{\text{out}}\cdot\mathbf{r}_{\text{screen}}} F(\mathbf{q}),$$

where $\mathbf{q} = \mathbf{k}_{\text{out}} - \mathbf{k}_{\text{in}}$. The measured intensity of the reflection will be square of this amplitude,

$$A^2 S^2 \left|F(\mathbf{q})\right|^2.$$

Friedel and Bijvoet Mates

For every reflection corresponding to a point $\mathbf{q}$ in the reciprocal space, there is another reflection of the same intensity at the opposite point $-\mathbf{q}$. This opposite reflection is known as the *Friedel mate* of the original reflection. This symmetry results from the mathematical fact that the density of electrons $f(\mathbf{r})$ at a position $\mathbf{r}$ is always a real number $f(\mathbf{r})$ is the inverse transform of its Fourier transform $F(\mathbf{q})$; however, such an inverse transform is a complex number in general. To ensure that $f(\mathbf{r})$ is real, the Fourier transform $F(\mathbf{q})$ must be such that the Friedel mates $F(-\mathbf{q})$ and $F(\mathbf{q})$ are complex conjugates of one another. Thus, $F(-\mathbf{q})$ has the same magnitude as $F(\mathbf{q})$ but they have the opposite phase, i.e., $\varphi(\mathbf{q}) =-\varphi(\mathbf{q})$.

$$F(-\mathbf{q}) = \left|F(-\mathbf{q})\right| e^{i\phi(-\mathbf{q})} = F^*(\mathbf{q}) = \left|F(\mathbf{q})\right| e^{-i\phi(\mathbf{q})}.$$

The equality of their magnitudes ensures that the Friedel mates have the same intensity $|F|^2$. This symmetry allows one to measure the full Fourier transform from only half the reciprocal space, e.g., by rotating the crystal slightly more than 180° instead of a full 360° revolution. In crystals with significant symmetry, even more reflections may have the same intensity (Bijvoet mates); in such cases, even less of the reciprocal space may need to be measured. In favorable cases of high symmetry, sometimes only 90° or even only 45° of data are required to completely explore the reciprocal space.

The Friedel-mate constraint can be derived from the definition of the inverse Fourier transform,

$$f(\mathbf{r}) = \int \frac{d\mathbf{q}}{(2\pi)^3} F(\mathbf{q}) e^{i\mathbf{q}\cdot\mathbf{r}} = \int \frac{d\mathbf{q}}{(2\pi)^3} \left|F(\mathbf{q})\right| e^{i\phi(\mathbf{q})} e^{i\mathbf{q}\cdot\mathbf{r}}.$$

Since Euler's formula states that $e^{ix} = \cos(x) + i\sin(x)$, the inverse Fourier transform can be separated into a sum of a purely real part and a purely imaginary part,

$$f(\mathbf{r}) = \int \frac{d\mathbf{q}}{(2\pi)^3} |F(\mathbf{q})| e^{i(\phi + \mathbf{q}\cdot\mathbf{r})}$$

$$= \int \frac{d\mathbf{q}}{(2\pi)^3} |F(\mathbf{q})| \cos(\phi + \mathbf{q}\cdot\mathbf{r}) + i\int \frac{d\mathbf{q}}{(2\pi)^3} |F(\mathbf{q})| \sin(\phi + \mathbf{q}\cdot\mathbf{r}) = I_{\cos} + iI_{\sin}.$$

The function $f(\mathbf{r})$ is real if and only if the second integral $I_{\sin}$ is zero for all values of $\mathbf{r}$. In turn, this is true if and only if the above constraint is satisfied,

$$I_{\sin} = \int \frac{d\mathbf{q}}{(2\pi)^3} |F(\mathbf{q})| \sin(\phi + \mathbf{q}\cdot\mathbf{r}) = \int \frac{d\mathbf{q}}{(2\pi)^3} |F(-\mathbf{q})| \sin(-\phi - \mathbf{q}\cdot\mathbf{r}) = -I_{\sin},$$

since $I_{\sin} = -I_{\sin}$ implies that $I_{\sin} = 0$.

Ewald's Sphere

Each X-ray diffraction image represents only a slice, a spherical slice of reciprocal space, as may be seen by the Ewald sphere construction. Both $\mathbf{k}_{out}$ and $\mathbf{k}_{in}$ have the same length, due to the elastic scattering, since the wavelength has not changed. Therefore, they may be represented as two radial vectors in a sphere in reciprocal space, which shows the values of **q** that are sampled in a given diffraction image. Since there is a slight spread in the incoming wavelengths of the incoming X-ray beam, the values of $|F(\mathbf{q})|$ can be measured only for **q** vectors located between the two spheres corresponding to those radii. Therefore, to obtain a full set of Fourier transform data, it is necessary to rotate the crystal through slightly more than 180°, or sometimes less if sufficient symmetry is present. A full 360° rotation is not needed because of a symmetry intrinsic to the Fourier transforms of real functions (such as the electron density), but "slightly more" than 180° is needed to cover all of reciprocal space within a given resolution because of the curvature of the Ewald sphere. In practice, the crystal is rocked by a small amount (0.25–1°) to incorporate reflections near the boundaries of the spherical Ewald's shells.

Patterson Function

A well-known result of Fourier transforms is the autocorrelation theorem, which states that the autocorrelation $c(\mathbf{r})$ of a function $f(\mathbf{r})$,

$$c(\mathbf{r}) = \int d\mathbf{x} f(\mathbf{x}) f(\mathbf{x}+\mathbf{r}) = \int \frac{d\mathbf{q}}{(2\pi)^3} C(\mathbf{q}) e^{i\mathbf{q}\cdot\mathbf{r}}$$

has a Fourier transform $C(\mathbf{q})$ that is the squared magnitude of $F(\mathbf{q})$:

$$C(\mathbf{q}) = |F(\mathbf{q})|^2.$$

Therefore, the autocorrelation function $c(\mathbf{r})$ of the electron density (also known as the *Patterson function*) can be computed directly from the reflection intensities, without computing the phases. In principle, this could be used to determine the crystal structure directly; however, it is difficult to realize in practice. The autocorrelation function corresponds to the distribution of vectors between atoms in the crystal; thus, a crystal of N atoms in its unit cell may have $N(N - 1)$ peaks in its Patterson function. Given the inevitable errors in measuring the intensities, and the mathematical difficulties of reconstructing atomic positions from the interatomic vectors, this technique is rarely used to solve structures, except for the simplest crystals.

Advantages of a Crystal

In principle, an atomic structure could be determined from applying X-ray scattering to non-crystalline samples, even to a single molecule. However, crystals offer a much stronger signal due to their periodicity. A crystalline sample is by definition periodic; a crystal is composed of many unit cells repeated indefinitely in three independent directions. Such periodic systems have a Fourier transform that is concentrated at periodically repeating points in reciprocal space known as *Bragg peaks*; the Bragg peaks correspond to the reflection spots observed in the diffraction image. Since the amplitude at these reflections grows linearly with the number N of scatterers, the observed *intensity* of these spots should grow quadratically, like N^2. In other words, using a crystal concentrates the weak scattering of the individual unit cells into a much more powerful, coherent reflection that can be observed above the noise. This is an example of constructive interference.

In a liquid, powder or amorphous sample, molecules within that sample are in random orientations. Such samples have a continuous Fourier spectrum that uniformly spreads its amplitude thereby reducing the measured signal intensity, as is observed in SAXS. More importantly, the orientational information is lost. Although theoretically possible, it is experimentally difficult to obtain atomic-resolution structures of complicated, asymmetric molecules from such rotationally averaged data. An intermediate case is fiber diffraction in which the subunits are arranged periodically in at least one dimension.

Applications of X-ray Diffraction

X-ray diffraction has wide and various applications in the chemical, biochemical, physical, material and mineralogical sciences. Laue claimed in 1937 that the technique "has extended the power of observing minute structure ten thousand times beyond that given us by the microscope". X-ray diffraction is analogous to a microscope with atomic-level resolution which shows the atoms and their electron distribution.

X-ray diffraction, electron diffraction, and neutron diffraction give information about the structure of matter, crystalline and non-crystalline, at the atomic and molecular level. In addition, these methods may be applied in the study of properties of all materials, inorganic, organic or biological. Due to the importance and variety of applications of diffraction studies of crystals, many Nobel Prizes have been awarded for such studies.

X-ray Method for Investigation of Drugs

X-ray diffraction has been used for the identification of antibiotic drugs such as: eight β-lactam (ampicillin sodium, penicillin G procaine, cefalexin, ampicillin trihydrate, benzathine penicillin, benzylpenicillin sodium, cefotaxime sodium, ceftriaxone sodium), three tetracycline (doxycycline hydrochloride, oxytetracycline dehydrate, tetracycline hydrochloride) and two macrolide (azithromycin, erythromycin estolate) antibiotic drugs. Each of these drugs has a unique X-Ray Diffraction (XRD) pattern that makes their identification possible.

X-ray Method for Investigation of Textile Fibers and Polymers

Forensic examination of any trace evidence is based upon Locard's exchange principle. This states that "every contact leaves a trace". In practice, even though a transfer of material has taken place, it may be impossible to detect, because the amount transferred is very small.

Textile fibers are a mixture of crystalline and amorphous substances. Therefore, the measurement of the degree of crystalline gives useful data in the characterization of fibers using X-ray diffractometry. It has been reported that X-ray diffraction was used to identify of a "crystalline" deposit which was found on a chair. The deposit was found to be amorphous, but the diffraction pattern present matched that of polymethylmethacrylate. Pyrolysis mass spectrometry later identified the deposit as polymethylcyanoacrylaon of Boin crystal parameters.

X-ray Method for Investigation of Bones

Heating or burning of bones causes recognizable changes in the bone mineral that can be detected using XRD techniques. During the first 15 minutes of heating at 500 °C or above, the bone crystals began to change. At higher temperatures, thickness and shape of crystals of bones appear stabilized, but when the samples were heated at a lower temperature or for a shorter period, XRD traces showed extreme changes in crystal parameters.

Integrated Circuits

X-ray diffraction has been demonstrated as a method for investigating the complex structure of integrated circuits.

Crystallographic Disorder

In X-ray crystallography, crystallographic disorder describes the cocrystallization of more than one rotamer, conformer, or isomer where the center of mass of each form is identical or unresolvable. As a consequence of disorder, the crystallographic solution is the sum of the various forms. In many cases, the components of the disorder are equally abundant, and, in other cases, the weighting coefficients for each component differ. Disorder can entail a pair or several components. Disorder usually arises when the forms are nearly equal in energy and the crystal lattice is sufficiently spacious to accommodate the various components.

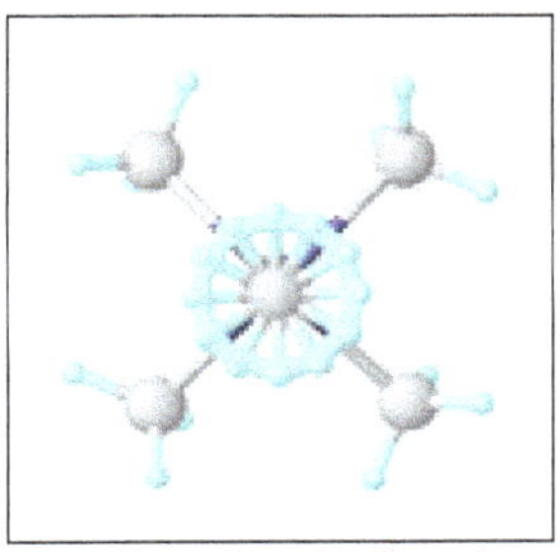

Structure of $Mo(CH_3)_5$ showing 4-fold disorder of one methyl group.

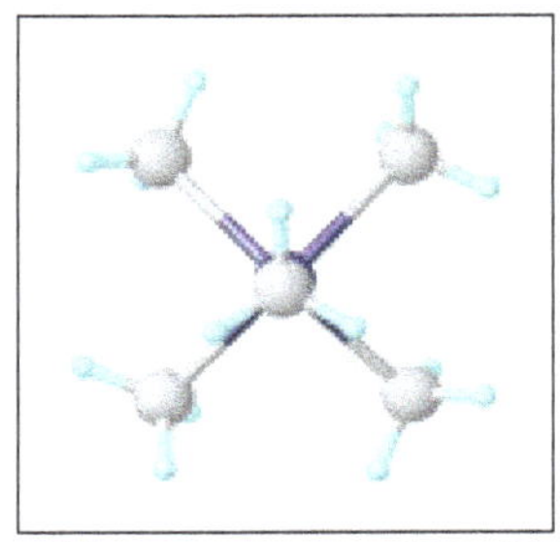

One of four rotamers of $Mo(CH_3)_5$.

Classification of Disorders

Dynamic Disorder

Atoms and molecules are never fixed. They are vibrating, and these movements are increasing with temperature. The so called Brownian motion has first been observed in 1827 by the Scottish botanist Robert Brown observing the unexplainable motion of small particles from plants held in an aqueous suspension. Later it could be shown that the description of this motion is a good model to describe random physical processes. In the case of molecules these movements are very fast, usually in the order of magnitude of 10^{12} and 10^{14} Hz. As exposure times with standard equipment in laboratories are between a few seconds and some minutes it is clear that these experiments cannot resolve these motions of the molecules. The refined structure therefore represents the average of all the possible positions of all the atoms being involved.

In the case of a standard structure determination the motion of the atoms is described as a harmonic oscillation of the atom around the coordinates that have been attributed to that atom. This atomic model is called the "independent atom model" or IAM model.

The electrons are assumed to be distributed spherically around the nucleus in this model, and the only deformations it describes are those caused by thermal motion. Its big advantage is that it is simple and uses only nine parameters to describe the electron distribution around the atom: three coordinates and six displacement parameters. Depending on the environment of a given atom and on the temperature the thermal motions can become very large. As the IAM model does not account for effects imposed by the chemical bonding it comes quickly to its limits. While the effective motion of the atoms is usually on some circular path it tries to describe the electron density of the vibrating atom using a linear harmonic motion resulting in residual electron density higher than desired. This situation is called dynamic disorder. It is usually sensitive to temperature and its effects can be decreased by lowering the ambient temperature of the data collection. Figure shows an example of a data collection carried out at a temperature too high to achieve good separation of the individual atomic sites.

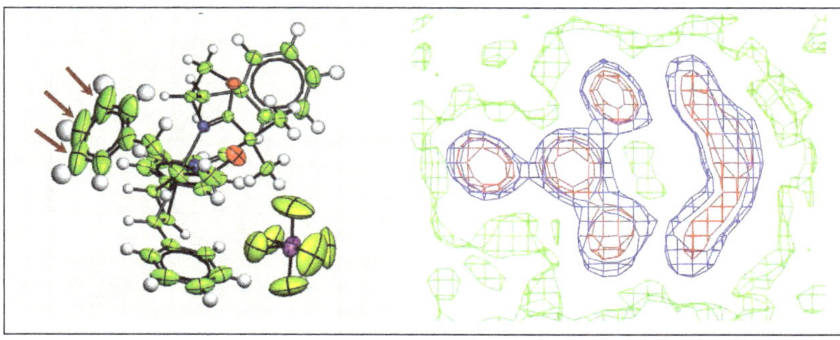

The structure of a Palladium complex measured at 250K. In particular the phenyl ring shown left is affected by dynamic disorder. The slant Fourier map through the plane defined by the ring shows how the electron density is spread over the three peripheral positions of the ring. The electron density for the three atoms indicated by the arrows on the left side shows no separation between the atoms.

Static Disorder

During the process of crystallization every molecule that is joining the crystal has to find the best position with lowest energy itself. It is possible that there is more than one way with comparably low energy to join the crystal. A peripheral group with rotational freedom may have similar space requirements in different orientations. There may be more than one possibility to interact with the local environment for a co-crystallizing solvent molecule, or the absence of interactions may give a molecule or molecular fragment the freedom to orient randomly.

As the structural fragments that vary are not explainable as the oscillation around a common position caused by thermal motion their contributions to the final diffraction pattern are different, even if only the resulting sum can be determined. This is called static disorder as the disordered parts will remain in their positions even if ambient conditions are changed. A dynamic disorder will be influenced if temperature is lowered. Decreasing the temperature in the case of a static disorder will of course also decrease the displacement parameters of the disordered part, but the disorder itself will not be affected.

This was the case for the complex shown in figure. The allyl group can bind to the Palladium atom in two orientations distinguished by the position of the central carbon atom pointing up or down respectively. In solution these two conformations are in equilibrium. By NMR studies in solution at different temperatures it could be observed that lowering the temperature changed the equilibrium in favor of one of the two conformations, and at about 200K the signals of the less frequent conformation disappeared completely. Crystals grown at room temperature were used for the structure determination. Repeating the experiment at different temperature showed no change in the ratio of the site occupancy factors of about 3:1 which illustrates nicely the static nature of the disorder in the crystal structure.

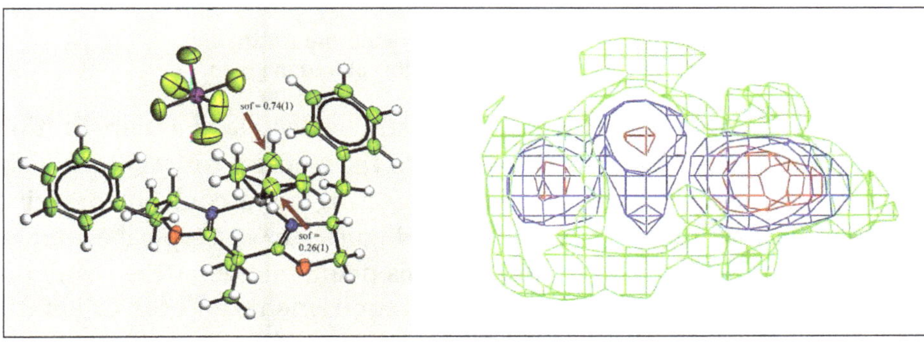

The structure of a Palladium complex measured at 250K. The allyl group is statically disordered, the site occupancy factors of the two positions of the central Carbon atom refined to 0.74(1) and 0.26(1), the two positions are marked by arrows. The Carbon atom with higher site occupancy can be seen well in the slant Fourier map and the position of the less occupied central carbon atom is separated from it.

In the absence of long range order there is no way of separating the molecules with different conformation and the result is that we see the variations of the structure as a superimposition of all conformations present. It is not entirely clear why in some cases there is long range order in the local variations and in others not. The speed of crystal growth, the pureness of the compound to crystallize and the ambient conditions during crystallization may be reasons that can be in favor or in disfavor of building long range order.

Substitutional Disorder

There is another case of disorder that can be observed mostly with inorganic materials like minerals. Different elements with similar binding properties can occupy equivalent places in a structure. This type of disorder is therefore called substitutional disorder.

In order to cope well with this type of disorder accurate knowledge of the composition is extremely helpful. Substitutional disorder is static by nature. Amongst the difficulties to refine such disorders is the fact that the bond length to the environment may vary resulting in bad displacement parameters. Figure shows the structure of the mineral Jentschite where Antimony can take the place of Arsenic.

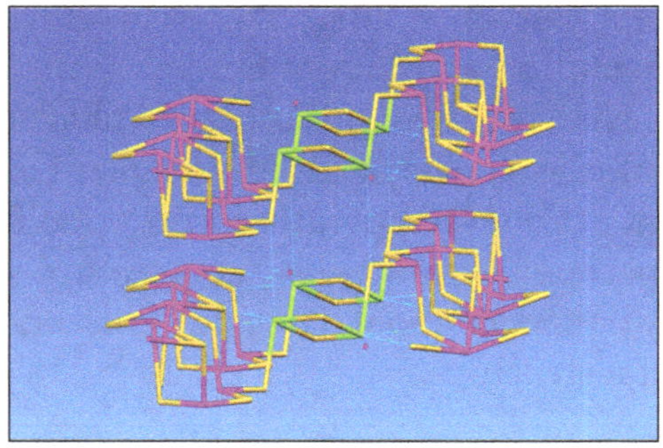

Jentschite, the substitution of Arsenic and Antimony can be observed at the atomic site colored in green.

There are examples in organic structures that could be described as substitutional disorder. The frequent case of a thiophene ring with the sulfur position being disordered between the two adjacent positions to the connecting bond could be described as a substitution of a Carbon atom by a Sulfur atom. Figure shows a typical example of this disorder. But it is probably better to describe this phenomenon as the result of effects taking place during crystallization. As the space requirements for both orientations are very similar and as there are no significant interactions guiding the crystallization process some of the thiophene rings will be oriented in one way while the rest will take the opposite orientation. Substitutional disorder suggests the substitution in the moment of the synthesis or crystal growth. Even though a Sulfur atom takes the place of a carbon atom and vice versa the different orientations of the thiophene ring are a result of the crystallization process and would therefore be rather classified as a static than a substitutional disorder.

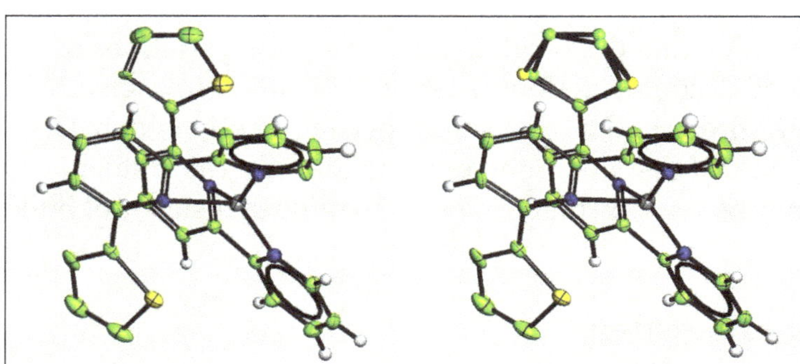

Disordered thiophene ring as seen after initial refinement and a disorder model with the original positions rotated around the bonding axis by 180 degrees. The second thiophene ring visible in the bottom of the image shows similar signs of the disorder described, and in fact it is possible to refine also this ring with a similar disorder model. The result is a second group of atoms with a site occupancy factor of about 5%, and not much can be said about the correctness of the model. 5% of a carbon atom is in the region of electron density where we only expect the residual electron density, therefore it is difficult to conclude if the assumption is correct or not.

Solvent Disorder

Disordered structures can also be classified on the basis of what part of the structure is disordered. As a rule of thumb it may be stated that the smaller the molecule that we want to look at the bigger the probability that it moves with temperature. Thus disorder in solvent molecules tends to be more frequent than disorder in bigger molecules. This problem is of particular interest and importance in macromolecular crystallography and in protein crystallography where it is quite frequent that up to 40% of a structure is simply filled with water molecules. The crystal is behaving like a sponge. Of course hydrogen bonds will build up, but they will, as long as the crystal is not cooled down, change their orientation and with this the whole network structure will be different.

Even though it is rare to find such high percentages of included solvent molecules in small molecule crystallography these structures exist as shown in figure. Solvent disorder sometimes presents tough problems to solve, and the difficulty is mostly due to the fact that disorder together with high thermal motion leads to flat electron density distributions where it is difficult to get started by assigning atom types to residual electron density maxima.

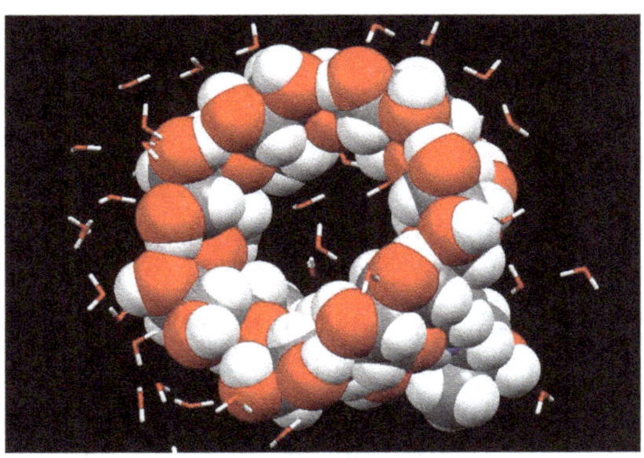

This structure contains 16 water molecules for one of b-cyclodextrine moiety.
The molecules shown are only those that could be localized in the density map.

As the refinement of disordered solvent molecules, in particular in macromolecular structures, may remain without good results special tools have been developed to find solutions that do not require a true atomic model for these solvent accessed areas.

Partial Fragment Disorder

If a larger molecule is affected by disorder then it is usually only one or more parts that are disordered while the bulk structure is ordered. Looking at the magnitudes of the displacement parameters it can be shown that usually U-values tend to be smaller in the center of the molecules where the degrees of freedom are restricted while peripheral groups make more frequent use of their degrees of freedom. But this observation is too general as there are examples of disorder being found in the very centre of molecules.

The fact that there are large parts of the structure that are ordered is usually a good indication that it is a true case of disorder that is under investigation. When large areas become suspect of being disordered it is increasingly difficult not only to handle the problem as a disorder problem, but it becomes also more and more likely that another reason causes that problem, for instance a wrong choice of the space group.

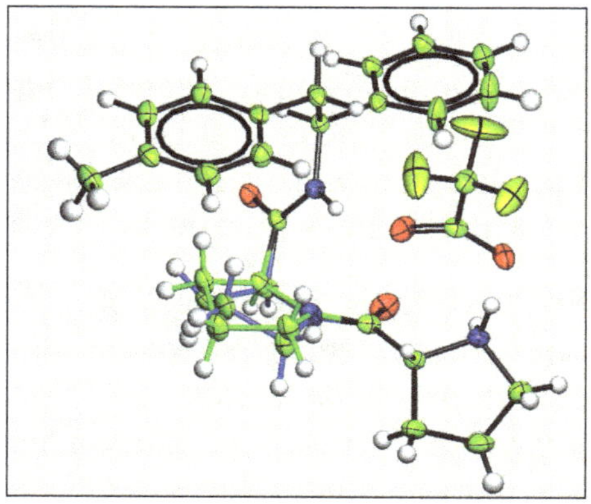

In this structure the disorder is located in the center of the molecule. The site occupancy factors of the two disordered parts refined to 0.509 (green) and 0.491(5) (blue).

Whole Molecule Disorder

It is in principle also possible that a whole structure is affected by disorder problems. However there is, as already mentioned, the strong suspicion that something else than disorder might be the true reason for the trouble and that limitations in the data collection equipment or in the crystal quality makes it impossible to see the true reason. In the hypothetic case of a modulated structure from a poorly diffracting crystal with satellites making up 1% of the total diffracting power of the crystal it is very easy that the investigator simply misses the extra observations, but without the additional data the modulation functions cannot be determined and what we see is apparently a disordered structure.

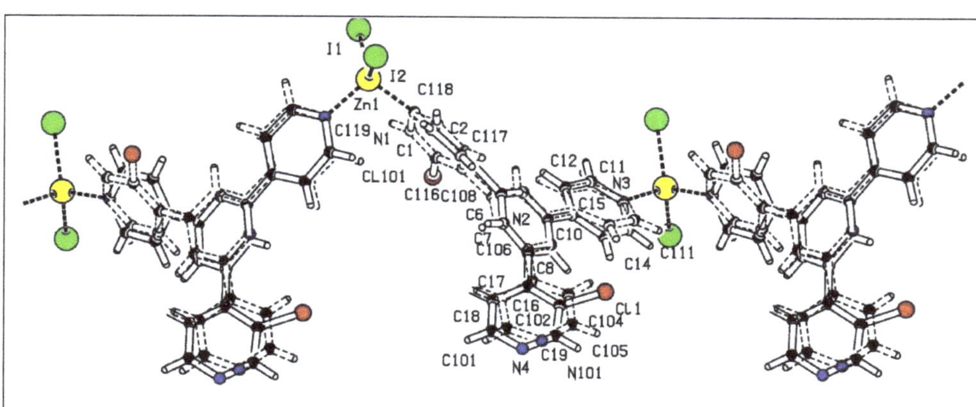

One entire ligand molecule is disordered in this structure. The less frequent arrangement of the ligand is shown with dashed bonds.

The structure with the highest percentage of disordered atoms observed in the crystallography lab in Basel was the structure of a metal organic framework built of metal ions and a bridging ligand. It is shown in figure. While the metal ions and the counter-ions were ordered the halogenated ligand showed two orientations throughout the whole structure.

Disordered Structures

A structure may look as if it would be disordered, but in fact the current description of the structure is incomplete and does not take into account other possible elements of ordering that may not be obvious to spot.

Structures with Z'>1

The static disorder has been described as the result of competing environmental conditions when the different possibilities to join the crystal are chosen randomly. It is possible that every possible arrangement is equally frequent, and their occurrence is alternating. If the structural features of the different contributors are similar and if the molecules are arranged in a pseudo-translational way, reflections along the direction of the alternation will be weakened in a systematic way. If the number of variants is n every reflection with the Miller index x*n along the involved axis will be observed while the ones in between will be weakened. The correct way of treating this case is to re-determine the unit cell constants taking into account weak reflections and to extract the intensity information again with the correct unit cell. The solution of such a structure will show more than one molecule in the asymmetric unit that cannot be related by crystallographic symmetry. The number of formula units in the asymmetric unit is also referred to as Z', and in this case we have a structure with a value of Z' > 1.

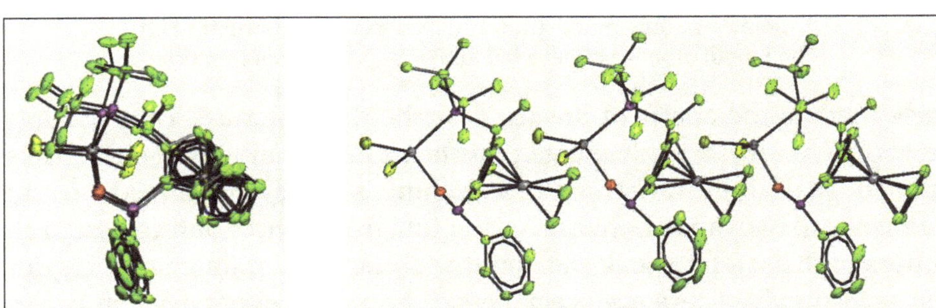

The unit cell of this structure was not determined correctly in a first run resulting in a structure that looked as if it would be disordered. Here the correct solution is shown. The apparently disordered structure appeared as a superimposition of all three slightly different molecules, the picture on the left hand side gives a good impression of this situation.

Even though the percentage of structures with Z' > 1 has probably always been equal it seems that modern equipment improves the chances of detecting the additional observations needed to determine these structures correctly. This apparently results in an increase of the number of structures with Z' >1. It is always recommended to test carefully if the pseudo symmetry is not matching missed higher space group symmetry.

Z'>1 and Commensurately Modulated Structures

Alternatively the same structure can be described as a modulated structure. The difference is that our model is made up of an average position for each atom and a modulation of the atoms around their corresponding average position. In the case of the structure shown in the figure above we would have the case of a commensurately modulated structure as the modulation has a fixed sequence of three states and returns to the origin. The original small unit cell would have to be taken for the integration of the frames and the weakened peaks in between the strong peaks would have to be integrated as so called satellites. In order to describe a modulation a new dimension has to be added to 3D space. For each direction in 3D space along which a modulation is detected a new dimension is created. The number of indices used to index reflections and satellites depends on the number of dimensions needed to describe both three dimensional space and the modulations and is restricted to three plus three dimensions as a maximum. The concept of modulated structures has been first described by P. M. de Wolff, and in this paper the term superspace is used to give the construction a name. Figure illustrates how the modulation works and why, if not treated, the structure may look as if it would be disordered.

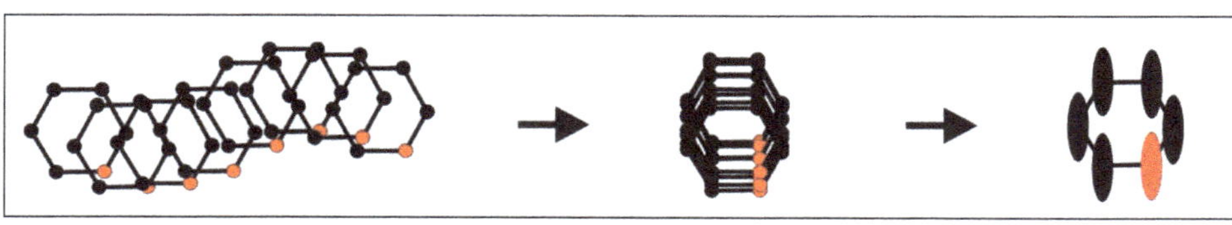

A simple case of a modulation and its effect on the structure if t he modulation is not detected and not taken care of properly.

In this case there is the choice of how to describe the structure. The description as a structure with Z' > 1 has the advantage that the usual techniques of structure solution can be applied. To describe the structure as commensurately modulated the integration of the frames needs advanced options of the integration program to assign the additional indices to the satellite peaks. Structure solution and refinement programs must be chosen accordingly. There are special space groups in superspace in order to take into account the additional dimensions, and the refinement of a modulated structure needs specialized skills and training. In the case of the example with Z' = 3 shown in figure where the refinement in three dimensional space is straight forward there is no need to choose the more complicated way of refining as a modulated structure. If Z' is growing the situation may change quickly.

If the molecules in a structure with Z' > 1 line up in almost perfect translational symmetry a part of the reflections along the axis in which the near translational symmetry is observed will be weakened. This can be the reason why these reflections are missed when indexing the crystal. These weak reflections can in extreme cases become too weak to be measured accurately. If the number of parameters to refine is growing while the number of observed intensities is small it may become difficult to refine such a structure in three dimensions. Choosing to refine the structure as commensurately modulate has a great advantage in this case. There is one molecule with its physical parameters to be refined, in addition there are the parameters of the atomic modulation functions to be refined for those atoms where a modulation is detected. This is in general a net advantage compared to refining three sets of possibly highly correlated parameters.

It is worth noting at this point that there are also examples where the description as commensurately modulated structures can be useful to get a better understanding of a structure or of a series of structures even if the single structures can be determined in three dimensional space. The advantage of such a description is clear as it makes it much easier to spot common features and differences in the structures when describing the differences as modulations of a common base structure.

Incommensurately Modulated Structures

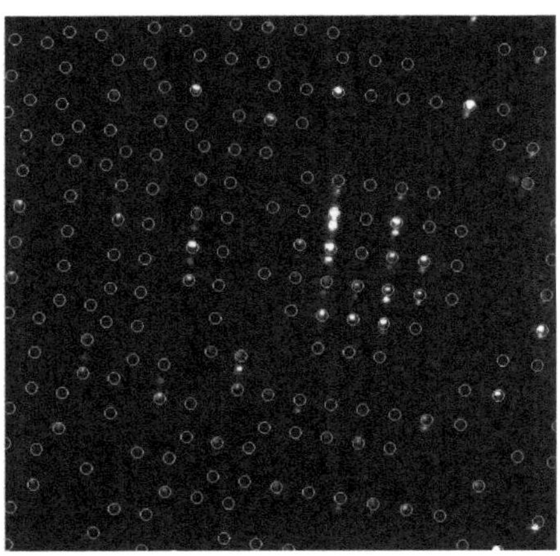

If the modulation function does not have a periodicity that falls together with a grid point of the average structure after a finite number of unit cells the structure is said to be incommensurately modulated. In this case there is no choice. The satellite peaks need to be indexed and the modulation needs to be determined for a proper description of the structure. Usually the modulations are not very big, and this means that the satellite peaks are weak in relation to the main peaks. The diffraction power of all satellites may be in the range of a few percent in relation to the main peaks, which can make it very difficult to observe them. The average structure may look weird in such a case and it will in most cases not be possible to find a good explanation. New detectors that are more sensitive will probably make the number of recognized modulated structures increase.

One frame out of the data collection of an incommensurately modulated structure. The white circles show where the main spots are expected, the satellites are located in equal distances from the main reflections. It is also possible to find incommensurately modulated structures where the satellite reflections are very well visible and cannot be overlooked. Figure is a frame of a data collection that has been carried out in the X-ray laboratory. The so called average structure could be solved while the refinement of the modulated structure is still to be finished.

Twinned Structures

Another problem related with symmetry may be the reason why a structure looks as if it would be disordered. But in this case, in contrast to the case with wrong space groups where too much symmetry is imposed on the structure, there is not enough symmetry present to describe the structure fully. This is the case with twinned structures, which are crystals in which there are differently oriented unit cells related by symmetry called the twin laws. Again we see a superimposition of two or more possible orientations of the same structure. The elegant thing with the twin law is that, once determined, it solves the problem by refining only one additional parameter for the twin element scale factors.

Sometimes the shape of the crystals reveals that the sample is composed of more than one lattice, some examples are shown in figure. This can also be a hint when doing structure determinations, but of course the absence of external features is no proof that the crystal is a single crystal.

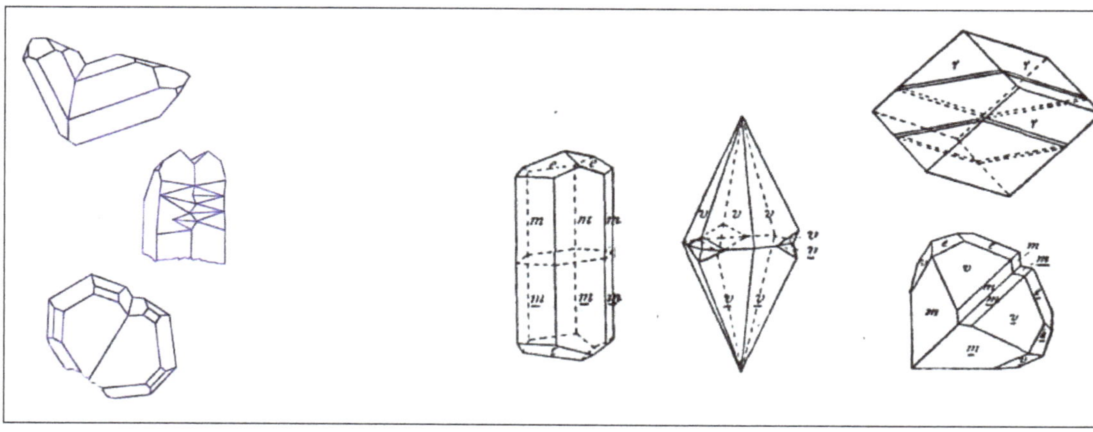

Drawings of twinned crystals of quartz and calcite.

Twinned crystals can be subdivided in merohedral and non-merohedral twins. While for merohedral or pseudo merohedral twins there is for each reflection in one component a reflection in the other component that overlaps with it, even though the reflections have their own miller indices in their respective unit cells, and the intensity they contribute to the final observation can be very different. Because of this such twins have an unexceptional diffraction pattern that shows no traces of the twinning. In spite of this problem it is possible to detect and resolve the twinning in later stages of the structure determination.

For non-merohedral twins, that is, twins where the contributions of the different components do not overlap, twinning may become obvious when the diffraction data fails to index. With persistence it may be possible to individually index the interpenetrating lattices. All components can then be indexed and integrated separately which gives a good chance to solve and refine the corresponding structure successfully.

Composite Structures

One type of structures that has also a different concept of ordering which goes beyond our intrinsic understanding of order in three dimensions should be mentioned. In the case of this type of structures there are host and guest molecules, and they have both their own periodic structure, but these two structures behave mostly independent from each other. Host and guest structure have each their own unit cell with independent Miller indices, both have their own space group and for each lattice the data must be extracted separately. The refinement program has to cope with the presence of the two datasets, and for each physical parameter it must be stated to which of the two substructures it is contributing.

Energy-dispersive X-ray Diffraction

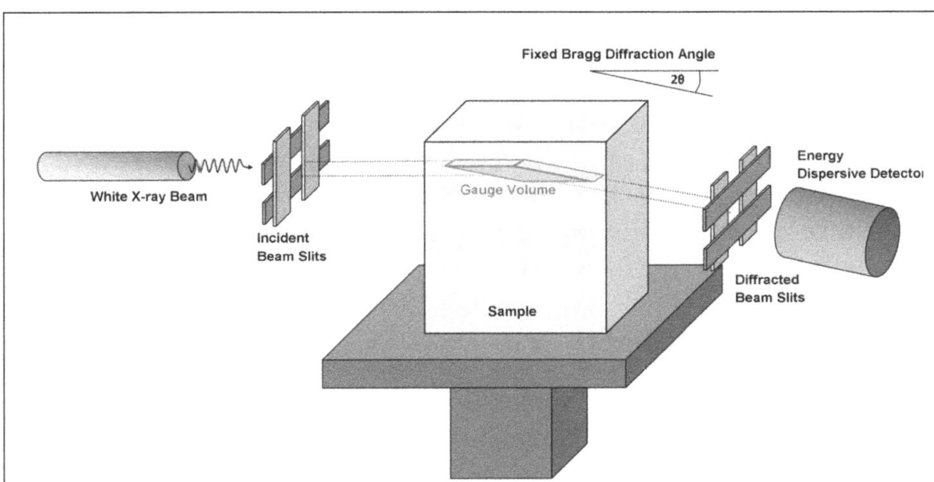

Schematic of a typical EDXRD experiment.

Energy-dispersive X-ray diffraction (EDXRD) is an analytical technique for characterizing materials. It differs from conventional X-ray diffraction by using polychromatic photons as the source and is usually operated at a fixed angle. With no need for a goniometer, EDXRD is able to collect full diffraction patterns very quickly. EDXRD is almost exclusively used with synchrotron radiation which allows for measurement within real engineering materials.

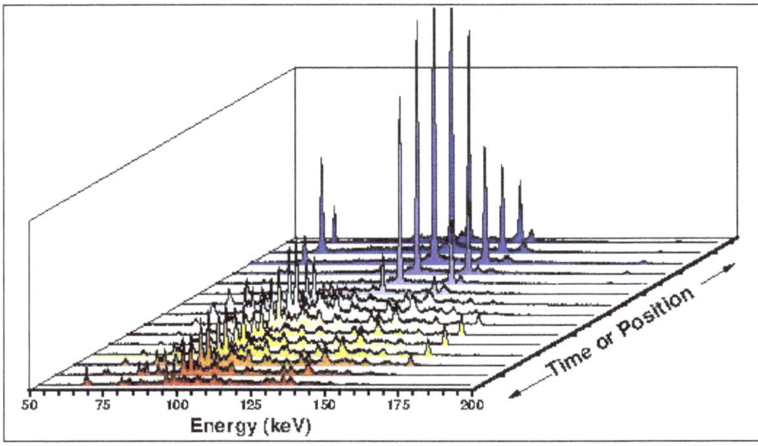

Sample data from an EDXRD experiment where synchrotron radiation is utilized. Spectra can be generated rapidly and as a function of position or time.

EDXRD was originally proposed independently by Buras et al. and Giessen and Gordon in 1968.

Advantages

The advantages of EDXRD are (1) it uses a fixed scattering angle, (2) it works directly in reciprocal space, (3) fast collection time, and (4) parallel data collection. The fixed scattering angle geometry makes EDXRD especially suitable for in situ studies in special environments (e.g. under very low or high temperatures and/or pressures). When the EDXRD method is used, only one entrance and one exit window are needed. The fixed scattering angle also allows for measurement of the diffraction vector directly. This allows for high-accuracy measurement of lattice parameters. It allows for rapid structure analysis and the ability to study materials that are unstable and only exist for short periods of time. Because the whole spectrum of diffracted radiation is obtained simultaneously, it enables parallel data collection studies where structural changes can be determined over time.

Clay Mineral X-ray Diffraction

Clay minerals are one of the most diverse minerals but all have a commonalty of crystal or grain sizes below 2 μm. Chemically clays are defined by crystal structure and chemical composition. Sometimes fine grain sediments are mistakenly described as clays, this is actually a description of the "clay-size fraction" rather than the mineralogy of the sediment. There are three crystallographic clay groups: platy clays (phyllosilicates), fibrous clay minerals, and amorphous clay. Phyllosilicates are the most abundant clays and are categorized based on the layering of a tetrahedral and an octahedral layer. For most clays, the octahedral layer is centered with Al^{3+}, Fe^{3+}, or $Mg(OH)_2$, but sometimes Zn^{2+}, Li^+, and Cr^{3+} can substitute as well. Si^{4+} is normally the center of the tetrahedral layer but Al^{3+} will often partially substitute and create a charge imbalance. Two-layer clays are composed of a tetrahedral layer and an octahedral layer (T-O) while three-layer clays contain an octahedral layer sandwiched by two tetrahedral layers (T-O-T). When substitution of Al^{3+} for Si^{4+} creates a charge imbalance, an interlayer cation will fill in between tetrahedral layers to balance the charge of the clay.

X-ray Diffraction and Clays

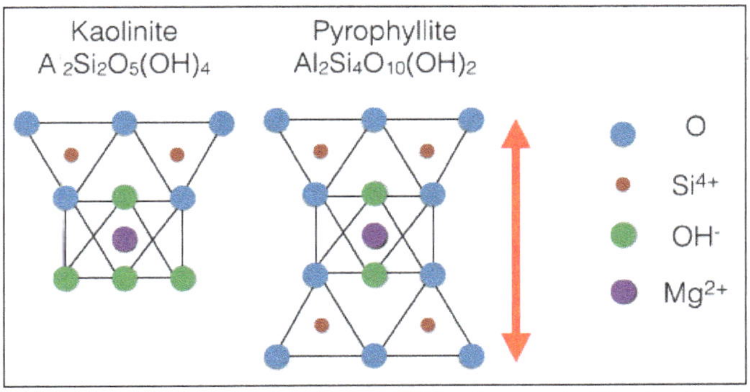

Structure of a two-layer clay, kaolinite (T-O) and pyrophyllite (T-O-T) with red arrow to represent the Z dimension. T-O-T clays are naturally larger in the Z direction because of the extra layer provided by the additional tetrahedral layer.

Typically, powder X-ray diffraction (XRD) is an average of randomly oriented micro-crystals that should equally represent all crystal orientation if a large enough sample is present. X-rays are directed at the sample while slowly rotated which produce a diffraction pattern which show intensity of x-rays collected at different angles. Randomly oriented XRD samples are not as useful for clay minerals because clays typically have similar X and Y dimensions. The Z dimension differs from clay to clay and is most diagnostic because the Z dimension represents the height of the tetrahedral-octahedral (T-O) or tetrahedral-octahedral-tetrahedral (T-O-T) layer. The Z dimension can increase or decrease because of substitution of the central cation in both the tetrahedral and octahedral layers. The presence and size of a charge balancing cation in the interlayer of T-O-T clays will also affect the Z dimension. Because of this, clay minerals are typically identified by preparing samples so that they are oriented to increase basal (00*l*) reflection . D positions are calculated using Bragg's law but because clay mineral analysis is one dimensional, *l* can substitute n, making the equation $l\lambda = 2d \sin\Theta$. When measuring the x-ray diffraction of clays, d is constant and λ is the known wavelength from the x-ray source, so the distance from one 00*l* peak to another is equal.

Identification of Clays using XRD

Basal reflections give d-spacing of the basal layer which represent the thickness of the silicate layers and the unit cell often contains multiple layers. Clay mineral peaks are can generally be distinguished by the width halfway up the peak (i.e. the full width at half maximum, FWHM). Well-defined crystalline minerals have sharp peaks while clays, which range from crystalline to noncrystalline, produce broad peaks with noticeable width on both sides. These broad peaks make it easy to pick out which peaks are contributed by clays. These peaks can be compared to known diffraction patterns for better identification but if some peaks are broader than others, it is likely that multiple clays are present. The Clay Mineral Society maintains a collection clays for the purpose of comparison to unknown clays. Because the majority of the clays available from the Clay Mineral Society are naturally formed, they can contain minerals other than the desired clay. Diffraction patterns calculated using theoretical methods do not generally match with experimental diffraction patterns, so using diffraction patterns from known samples to help identify a clay is preferable to calculation. Some minerals can be eliminated from identification using background information or prior analysis.

Well-crystallized and pure samples are ideal for x-ray diffraction, but this is rarely the case for clay. Clay minerals are almost always mixed with very small amounts of nonclay minerals which can produce intense peaks, even when there is very little of the sample is not a clay. If additional minerals are known to be presents, attempts should be made to separate clays from nonclays otherwise additional peaks should be expected. Some common minerals associated with naturally occurring clays are; quartz, feldspars, zeolites, and carbonates organic matter is sometimes present. Synthesis of clays can reduce the presence of some of these associated materials but does not guarantee pure samples as quartz or other associated materials are still commonly produced alongside synthetic clays.

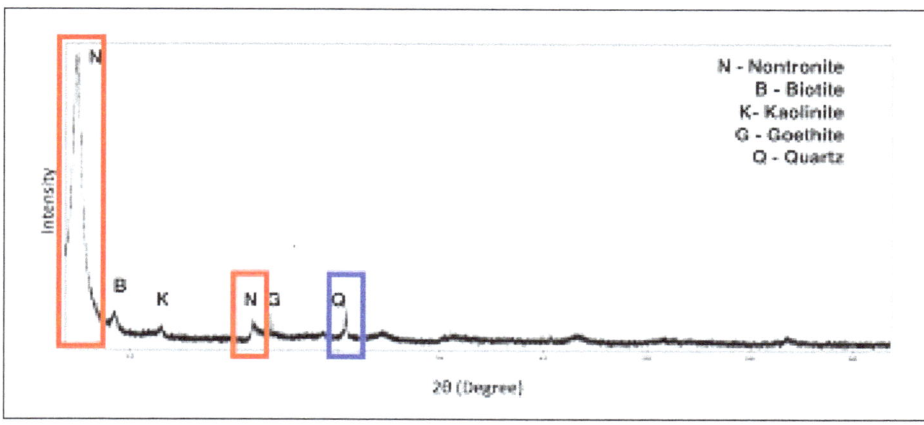

X-ray diffraction pattern of a Nontronite standard purchased from the clay mineral society. Red boxes represent a wide peaks attributed to nontronite. Blue box represents a sharp peak attributed to quartz present in the standard.

Mix Layered Clay Minerals

Mixed–layering, interlaying, and interstratification are all terms that refer to clay minerals that form with two or more types of clays that with intergrown layering. Mixed-layering does not refer to clays that have been physically mixed. Mixed-layering in clays make add difficulty to interpretation so multiple analysis are usually necessary. Two component clays are most common with multicomponent clays containing more than two components are very rare. The entire diffraction pattern contributes to the identification and peaks should be considered as a whole rather than individually.

Mixed layer clays with two equal components (50% of each clay) are the easiest to identify. These clays are thought of as one, non-mixed clay with a 001 spacing that is equal to the sum of the 001 spacing of both components. Some commonly occurring 50/50 mixed-layer clays are even given unique names such as dozyite, a Serpentine/Chlorite. Mixed clays that have unequal components with random stacking produce aperiodic 00*l* diffraction patterns known as irrational patterns. The coefficient of variation (CV) is the percent standard deviation of the average of d(001) calculated from various reflections. If CV is less than 0.75% then the mineral is given a unique name. If CV is greater than 0.75% then mixed-layered nomenclature is used.

Preparation for Clay Mineral X-ray Diffraction

Clays should be separated from the nonclay minerals to reduce interference of 00*l* peaks. Nonclay minerals can usually be separated by sieving samples at a small enough mesh. Samples should be lightly crushed but not pulverized because nonclay minerals will be reduced to clay sizes and become impossible to separate from the sample. Lightly crushing breaks apart the soft clays while keeping harder nonclays intact for easier removal.

Samples should be as homogeneous as possible, both in grain size and composition before mounting them for X-ray diffraction and long, flat, and thick samples are ideal. Four methods are commonly used for sample preparation and vary in difficulty and appropriateness of use.

Glass Slide Method

Easiest and fastest of the four commonly used methods but also the least accurate. A glass microscope slide is covered with a suspension of sample in water then placed in an oven at 90 °C and left to dry. For some samples, drying at temperatures this high can damage the clays. In that case, drying at room temperature is an option but will require more time. Orientation is usually fair and particles are segregated with the finest particles toward the top. This method produces thin films which provide inaccurate diffraction intensities at moderate and high angles.

Smear Method

This is a quick method that is good identifying bulk sample constituents. The sample is crushed with a mortar and pestle until the powder is able to be brushed onto a glass slide. The powder is then mixed with a few drops of a dispersant solution, usually ethanol but others are available, and spread evenly over the slide. Both large and small grain size fractions can utilize this method.

Filter Membrane Peel Technique

This technique prevents size segregation by using either quick filtration or rapid stirring to overcome settling velocities. Sample is poured into a vacuum filter apparatus and filtered quickly but some liquid is left remaining so that air is not drawn though the sample, the remaining is liquid is then decanted. The damp sample is then inverted onto a glass slide and the filter paper is removed. Fast filtration allows a representative particle size to collect on the filter paper which is then inverted and exposed when mounted on a slide.

Centrifuged Porous Plate

Produces the best diffraction patterns out of the four most common methods but requires the most skill and is most time consuming. Upon completion, samples have thick aggregates and preferred orientation. A special apparatus designed to hold a porous ceramic plate is placed into a centrifuge container and filled with suspended sample. Centrifuging forces the liquid through the porous plate leaving the sample behind to be dried below 100 °C. An advantage of this method is that exchangeable cations can be removed by passing a chloride solution through the plate once the sample has been dried. Exchanging cations can be useful when establishing peaks for standards with variable interlayer cations. For example, nontronite has an interlayer which can contain both calcium and sodium. If an unknown sample was suspected to only contain one of these cations, a more accurate standard could be prepared by exchanging the undesired cation.

Serial Femtosecond Crystallography

Serial femtosecond crystallography (SFX) is a form of X-ray crystallography developed for use at X-ray free-electron lasers (XFELs). Single pulses at free-electron lasers are bright enough to generate resolvable Bragg diffraction from sub-micron crystals. However, these pulses also destroy the crystals, meaning that a full data set involves collecting diffraction from many crystals. This method of data collection is referred to as *serial*, referencing a row of crystals streaming across the X-ray beam, one at a time.

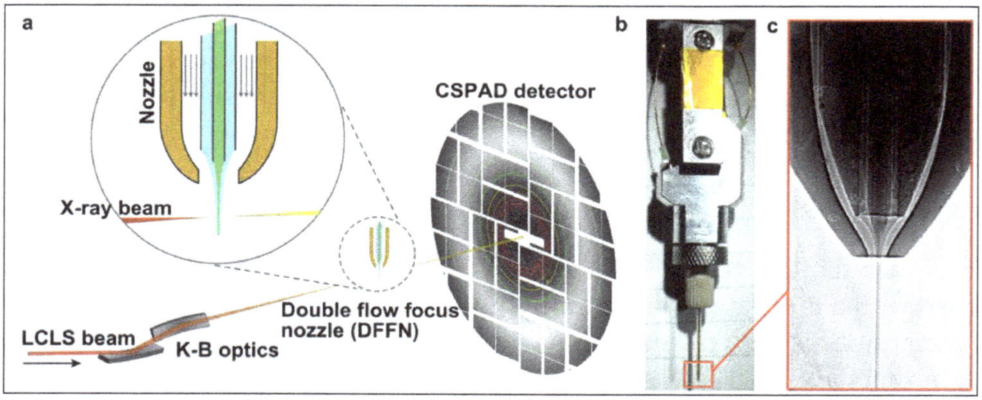

Serial Femtosecond Crystallography (SFX) schematic.

While the idea of serial crystallography had been proposed earlier, it was first demonstrated with XFELs by Chapman et al. at the Linac Coherent Light Source (LCLS) in 2011. This method has since been extended to solve unknown structures, perform time-resolved experiments, and later even brought back to synchrotron X-ray sources.

Methods

In comparison to conventional crystallography, where a single (relatively large) crystal is rotated in order to collect a 3D data set, some additional methods have to be developed to measure in the *serial* mode. First, a method is required to efficiently stream crystals across the beam focus. The other major difference is in the data analysis pipeline. Here, each crystal is in a random, unknown orientation which must be computationally determined before the diffraction patterns from all the crystals can be merged into a set of 3D *hkℓ* intensities.

Sample Delivery

The first sample delivery system used for this technique was the Gas Dynamic Virtual Nozzle (GDVN) which generates a liquid jet in vacuum (accelerated by a concentric helium gas stream) containing crystals. Since then, many other methods have been successfully demonstrated at both XFELs and synchrotron sources.

- Gas Dynamic Virtual Nozzle (GDVN): Low background scattering, but high sample consumption. Only method available for high repetition rate sources.
- Lipidic Cubic Phase (LCP) injector: Low sample consumption, with relatively high background. Specially suited for membrane proteins.
- Other viscous delivery media: Similar to LCP, low sample consumption with high background.
- Fixed target scanning systems (wide variety of systems have been used with different features, with standard crystal loops, or silicon chips): Low sample consumption, background depends on system, mechanically complex.
- Tape drive (crystals auto-pipetted onto a Kapton tape and brought to X-ray focus): Similar to fixed target systems, except with fewer moving parts.

Data Analysis

In order to recover a 3D structure from the individual diffraction patterns, they must be oriented, scaled and merged to generate a list of *hkℓ* intensities. These intensities can then be passed to standard crystallographic phasing and refinement programs. The first experiments only oriented the patterns and obtained accurate intensity values by averaging over a large number of crystals (> 100,000).

Small-angle X-ray Scattering

Small-angle X-ray scattering (SAXS) is a small-angle scattering technique by which nanoscale density differences in a sample can be quantified. This means that it can determine nanoparticle size distributions, resolve the size and shape of (monodisperse) macromolecules, determine pore sizes, characteristic distances of partially ordered materials, and much more. This is achieved by analyzing the elastic scattering behaviour of X-rays when travelling through the material, recording their scattering at small angles (typically 0.1 - 10°, hence the "Small-angle" in its name). It belongs to the family of small-angle scattering (SAS) techniques along with small-angle neutron scattering, and is typically done using hard X-rays with a wavelength of 0.07 - 0.2 nm. Depending on the angular range in which a clear scattering signal can be recorded, SAXS is capable of delivering structural information of dimensions between 1 and 100 nm, and of repeat distances in partially ordered systems of up to 150 nm. USAXS (ultra-small angle X-ray scattering) can resolve even larger dimensions, as the smaller the recorded angle, the larger the object dimensions that are probed.

SAXS and USAXS belong to a family of X-ray scattering techniques that are used in the characterization of materials. In the case of biological macromolecules such as proteins, the advantage of SAXS over crystallography is that a crystalline sample is not needed. Furthermore, the properties of SAXS allow investigation of conformational diversity in these molecules. Nuclear magnetic resonance spectroscopy methods encounter problems with macromolecules of higher molecular mass (> 30-40 kDa). However, owing to the random orientation of dissolved or partially ordered molecules, the spatial averaging leads to a loss of information in SAXS compared to crystallography.

Applications

SAXS is used for the determination of the microscale or nanoscale structure of particle systems in terms of such parameters as averaged particle sizes, shapes, distribution, and surface-to-volume ratio. The materials can be solid or liquid and they can contain solid, liquid or gaseous domains (so-called particles) of the same or another material in any combination. Not only particles, but also the structure of ordered systems like lamellae, and fractal-like materials can be studied. The method is accurate, non-destructive and usually requires only a minimum of sample preparation. Applications are very broad and include colloids of all types, metals, cement, oil, polymers, plastics, proteins, foods and pharmaceuticals and can be found in research as well as in quality control. The X-ray source can be a laboratory source or synchrotron light which provides a higher X-ray flux.

SAXS Instruments

In an SAXS instrument a monochromatic beam of X-rays is brought to a sample from which some of the X-rays scatter, while most simply go through the sample without interacting with it. The scattered X-rays form a scattering pattern which is then detected at a detector which is typically a 2-dimensional flat X-ray detector situated behind the sample perpendicular to the direction of the primary beam that initially hit the sample. The scattering pattern contains the information on the structure of the sample. The major problem that must be overcome in SAXS instrumentation is the separation of the weak scattered intensity from the strong main beam. The smaller the desired angle, the more difficult this becomes. The problem is comparable to one encountered when trying to observe a weakly radiant object close to the sun, like the sun's corona. Only if the moon blocks out the main light source does the corona become visible. Likewise, in SAXS the non-scattered beam that merely travels through the sample must be blocked, *without* blocking the closely adjacent scattered radiation. Most available X-ray sources produce *divergent* beams and this compounds the problem. In principle the problem could be overcome by *focusing* the beam, but this is not easy when dealing with X-rays and was previously not done except on synchrotrons where large bent mirrors can be used. This is why most laboratory small angle devices rely on collimation instead. Laboratory SAXS instruments can be divided into two main groups: point-collimation and line-collimation instruments.

Point-collimation Instruments

Point-collimation instruments have pinholes that shape the X-ray beam to a small circular or elliptical spot that illuminates the sample. Thus the scattering is centro-symmetrically distributed around the primary X-ray beam and the scattering pattern in the detection plane consists of circles around the primary beam. Owing to the small illuminated sample volume and the wastefulness of the collimation process — only those photons are allowed to pass that happen to fly in the right direction — the scattered intensity is small and therefore the measurement time is in the order of hours or days in case of very weak scatterers. If focusing optics like bent mirrors or bent monochromator crystals or collimating and monochromating optics like multilayers are used, measurement time can be greatly reduced. Point-collimation allows the orientation of non-isotropic systems (fibres, sheared liquids) to be determined.

Line-collimation Instruments

Line-collimation instruments restrict the beam only in one dimension (rather than two as for point collimation) so that the beam cross-section is a long but narrow line. The illuminated sample volume is much larger compared to point-collimation and the scattered intensity at the same flux density is proportionally larger. Thus measuring times with line-collimation SAXS instruments are much shorter compared to point-collimation and are in the range of minutes. A disadvantage is that the recorded pattern is essentially an integrated superposition (a self-convolution) of many adjacent pinhole patterns. The resulting smearing can be easily removed using model-free algorithms or deconvolution methods based on Fourier transformation, but only if the system is isotropic. Line collimation is of great benefit for any isotropic nanostructured materials, e.g. proteins, surfactants, particle dispersion and emulsions.

SAXS Instrument Manufacturers

SAXS instrument manufacturers include Anton Paar, Austria; Bruker AXS, Germany;

Hecus X-Ray Systems Graz, Austria; Malvern Panalytical. the Netherlands, Rigaku Corporation, Japan; Xenocs, France; and Saxslab, Denmark.

Wide-angle X-ray Scattering

In X-ray crystallography, wide-angle X-ray scattering (WAXS) or wide-angle X-ray diffraction (WAXD) is the analysis of Bragg peaks scattered to wide angles, which (by Bragg's law) are caused by sub-nanometer-sized structures. It is an X-ray-diffraction method and commonly used to determine the crystalline structure of polymers.

Wide-angle X-ray scattering is similar to small-angle X-ray scattering (SAXS) except the distance from sample to the detector is shorter and thus diffraction maxima at larger angles are observed. Depending on the measurement instrument used it is possible to perform both WAXS and SAXS in a single run (small- and wide-angle scattering, SWAXS).

Applications

The WAXS technique is used to determine of degree of crystallinity of polymer samples. It can also be used to determine the chemical composition or phase composition of a film, the texture of a film (preferred alignment of crystallites), the crystallite size and presence of film stress. As with other diffraction methods, the sample is scanned in a wide-angle X-ray goniometer, and the scattering intensity is plotted as a function of the 2θ angle.

X-ray diffraction is a non destructive method of characterization of solid materials. When X-rays are directed at solids they scatter in predictable patterns based on the internal structure of the solid. A crystalline solid consists of regularly spaced atoms (electrons) that can be described by imaginary planes. The distance between these planes is called the d-spacing.

The intensity of the d-space pattern is directly proportional to the number of electrons (atoms) in the imaginary planes. Every crystalline solid has a unique pattern of d-spacings (known as the powder pattern), which is a fingerprint for that solid. Solids with the same chemical composition but different phases can be identified by their pattern of d-spacings.

Flack Parameter

In X-ray crystallography, the Flack parameter is a factor used to estimate the absolute configuration of a structural model determined by single-crystal structure analysis.

In this approach, one determines the absolute structure of a noncentrosymmetric crystal. The processes used to decide the absolute structure use the anomalous dispersion effect. If atomic scattering factors did not have imaginary parts, the Friedel pairs would have exactly the same amplitudes (i.e., the scattering intensity $|F(hkl)|^2$ from crystal plane (h k l) is equal to $|F(-h-k-l)|^2$).

However, atomic scattering factors have imaginary parts due to the anomalous dispersion effect, and Friedel's law is broken by this effect.

There are several ways to determine the absolute structure by X-ray crystallography. For example, a comparison of the intensities of Bijvoet pairs or of the R-factors for the two possible structures can suggest the correct absolute structure. One of the more powerful and simple approaches is using the Flack parameter, because this single parameter clearly indicates the absolute structure.

The Flack parameter is calculated during the structural refinement using the equation given below:

$$I(hkl) = (1-x)|F(hkl)|^2 + x|F(-h-k-l)|^2$$

where x is the Flack parameter, I is the square of the scaled observed structure factor and F is the calculated structure factor.

By determining x for all data, x is usually found to be between 0 and 1. If the value is near 0, with a small standard uncertainty, the absolute structure given by the structure refinement is likely correct, and if the value is near 1, then the inverted structure is likely correct. If the value is near 0.5, the crystal may be racemic or twinned. The technique is most effective when the crystal contains both lighter and heavier atoms. Light atoms usually show only a small anomalous dispersion effect. In this case, the Flack parameter can refine to a physically unrealistic value (less than 0 or greater than 1) and has no meaning.

This parameter, introduced by H. D. Flack became one of a standard set of values being checked for structures with noncentrosymmetric space groups.

Multipole Density Formalism

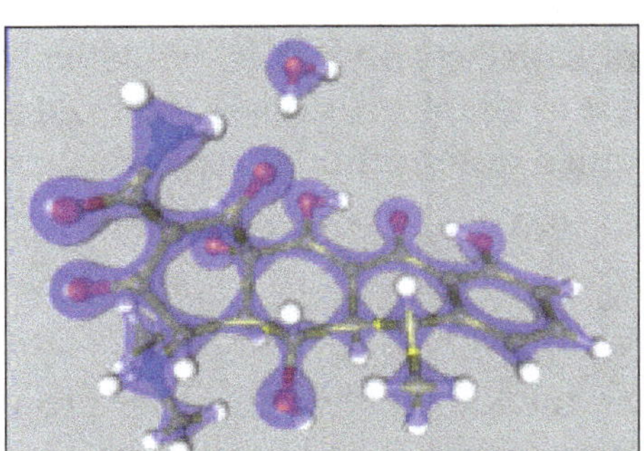

Isosurface of static electron density for doxycycline resulting from the Multipole Model refinement at non-target $1.5\,\mathring{A}/e^3$ level.

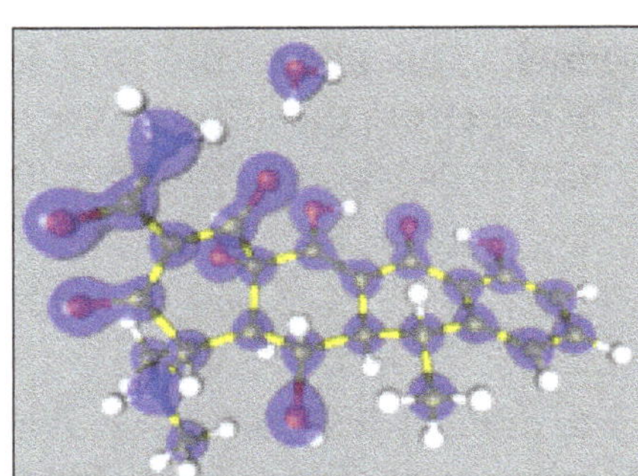

Equivalent isosurface resulting from the Independent Atom Model. Notice the lack of covalent bonds' electrons.

The Multipole Density Formalism (also referred to as Hansen-Coppens Formalism) is an X-ray crystallography method of electron density modelling proposed by Niels K.

Hansen and Philip Coppens in 1978. Unlike the commonly used Independent Atom Model, the Hansen-Coppens Formalism presents an aspherical approach, allowing one to model the electron distribution around a nucleus separately in different directions and therefore describe numerous chemical features of a molecule inside the unit cell of an examined crystal in detail.

Independent Atom Model

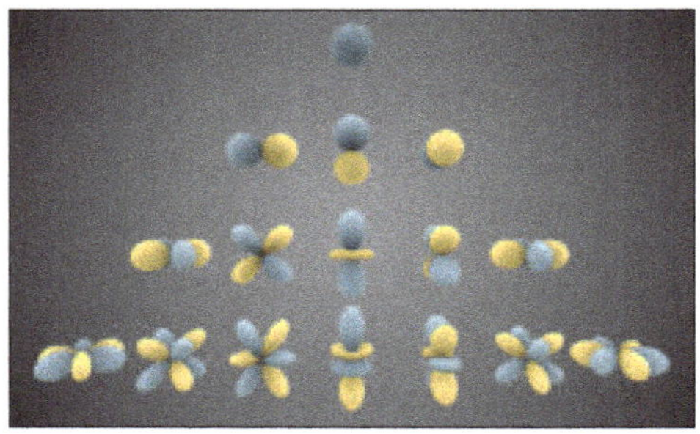

Visual representations of a few real spherical harmonics used in Multipole Density Formalism. The distance of the surface from the origin is proportional to the value for given angles, while the color denotes the sign (blue for positive, yellow for negative).

The Independent Atom Model (abbreviated to IAM), upon which the Multipole Model is based, is a method of charge density modelling. It relies on an assumption that electron distribution around the atom is isotropic, and that therefore charge density is dependent only on the distance from a nucleus. The choice of the radial function used to describe this electron density is arbitrary, granted that its value at the origin is finite. In practice either Gaussian- or Slater-type 1s-orbital functions are used.

Due to its simplistic approach, this method provides a straightforward model that requires no additional parameters (other than positional and Debye–Waller factors) to be refined. This allows the IAM to perform satisfactorily while a relatively low amount of data from the diffraction experiment is available. However, the fixed shape of the singular basis function prevents any detailed description of aspherical atomic features.

Kappa Formalism

In order to adjust some valence shell parameters, the Kappa formalism was proposed. It introduces two additional refineable parameters: an outer shell population (denoted as P_{val}) and its expansion/contraction (κ). Therefore, the electron density is formulated as:

$$\rho_{\text{atom}} = \rho_{\text{core}} + \rho'_{\text{valence}}(\kappa \text{r}) = \rho_{\text{core}} + \text{P}_{\text{val}}\kappa^{3}\rho_{\text{val}}(\kappa \text{r})$$

While P_{val}, being responsible for the charge flow part, is linearly coupled with partial charge, the normalised κ parameter scales radial coordinate r non-target. Therefore, lowering the κ parameter results in expansion of the outer shell and, conversely, raising it results in contraction. Although the Kappa formalism is still, strictly speaking, a spherical method, it is an important step towards understanding modern approaches as it allows one to distinguish chemically different atoms of the same element.

Multipole Description

In the multipole model description, the charge density around a nucleus is given by the following equation:

$$\rho_{atom}(r,\theta,\phi) = P_{core}\rho_{core}(r) + P_{val}\kappa^3\rho_{val}(\kappa r) + \sum_{l=0}^{l_{max}} \kappa'^3 R_l(\kappa' r) \sum_{m=-l}^{l} P_{lm\pm} Y_{lm\pm}(\theta,\phi)$$

The spherical part remains almost indistinguishable from the Kappa formalism, the only difference being one parameter corresponding to the population of the inner shell. The real strength of the Hansen-Coppens formalism lies in the right, deformational part of the equation. Here non-target κ' fulfils a role similar to non-target κ in the Kappa formalism (expansion/contraction of the aspherical part), whereas individual Non-target R_l are fixed spherical functions, analogous to non-target ρ. Spherical harmonics Non-target Y_{lm} (each with its populational parameter Non-target P_{lm}) are, however, introduced to simulate the electrically anisotropic charge distribution.

In this approach, a fixed coordinate system for each atom needs to be applied. Although at first glance it seems practical to arbitrarily and indiscriminately make it contingent on the unit cell for all atoms present, it is far more beneficial to assign each atom its own local coordinates, which allows for focusing on hybridisation-specific interactions. While the singular sigma bond of the hydrogen can be described well using certain z-parallel pseudoorbitals, xy-plane oriented multipoles with a 3-fold rotational symmetry will prove more beneficial for flat aromatic structures.

Applications

The primary advantage of the Hansen-Coppens formalism is its ability to free the model from spherical restraints and describe the surroundings of a nucleus far more accurately. In this way it becomes possible to examine some molecular features which would normally be only roughly approximated or completely ignored.

Hydrogen Positioning

X-ray crystallography allows the researcher to precisely determine the position of peak electron density and to reason about the placement of nuclei based on this information. This approach works without any problems for heavy (non-hydrogen) atoms, whose inner shell electrons contribute to the density function to a far greater degree then outer shell electrons.

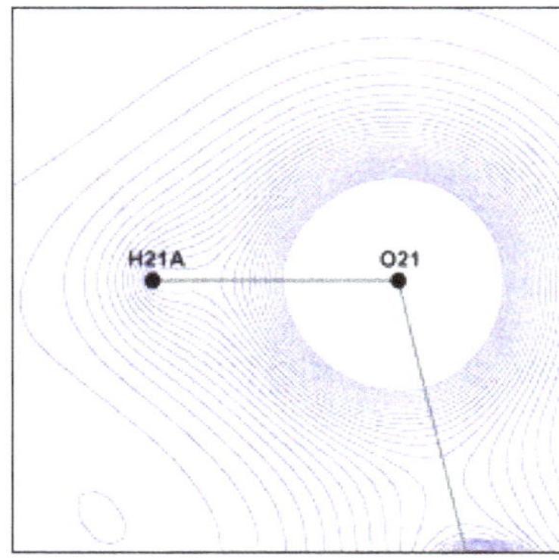

Electron density isosurface map around a covalent bond modelled with the Multipole Model, with populational parameters taken from the ELMAM2 database. Note the elongated high-density area next to the hydrogen atom, pointing in the direction of oxygen.

However, hydrogen atoms possess a feature unique among all the elements - they possess exactly one electron, which additionally is located on their valence shell and therefore is involved in creating strong covalent bonds with atoms of various other elements. While a bond is forming, the maximum of the electron density function moves significantly away from the nucleus and towards the other atom. This prevents any spherical approach from determining hydrogen position correctly by itself. Therefore, usually the hydrogen position is estimated basing on neutron crystallography data for similar molecules, or it is not modelled at all in the case of low-quality diffraction data.

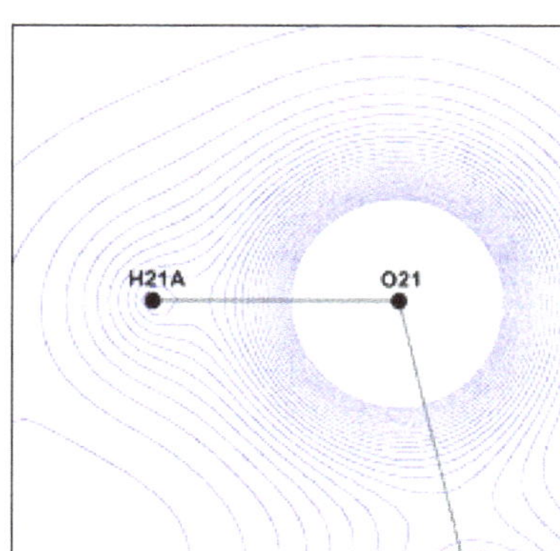

Electron density isosurface map around a covalent bond modelled with the Independent Atom Model in the same scale. Nucleus-centred functions impose lower charge density on the bond path.

It is possible (albeit disputable) to freely refine hydrogen atoms' positions using the Hansen-Coppens formalism, after releasing the bond lengths from any restraints derived from neutron measurements. The bonding orbital simulated with adequate multipoles describes the density distribution neatly while preserving believable bond lengths. It may be worth approximating hydrogen atoms' anisotropic displacement parameters, e.g. using SHADE, before introducing the formalism and, possibly, discarding bond distance constraints.

Bonding Modeling

In order to analyse the length and strength of various interactions within the molecule, Richard Bader's "Atoms in molecules" theorem may be applied. Due to the complex description of the electron field provided by this aspherical model, it becomes possible to establish realistic bond paths between interacting atoms as well as to find and characterise their critical points. Deeper insight into this data yields useful information about bond strength, type, polarity or ellipticity, and when compared with other molecules brings greater understanding about the actual electron structure of the examined compound.

Charge Flow

Due to the fact that for each multipole of every atom its population is being refined independently, individual charges will rarely be integers. In real cases, electron density flows freely through the molecule and is not bound by any restrictions resulting from the outdated Bohr atom model and found in IAM. Therefore, through e.g. an accurate Bader analysis, net atomic charges may be estimated, which again is beneficial for deepening the understanding of systems under investigation.

Drawbacks and Limitations

Although the Multipole Formalism is a simple and straightforward alternative means of structure refinement, it is definitely not flawless. While usually for each atom either three or nine parameters are to be refined, depending on whether an anisotropic displacement is being taken into account or not, a full multipole description of heavy atoms belonging to the fourth and subsequent periods (such as chlorine, iron or bromine) requires refinement of up to 37 parameters. This proves problematic for any crystals possessing large asymmetric units (especially macromolecular compounds) and renders a refinement using the Hansen-Coppens Formalism unachievable for low-quality data with an unsatisfactory ratio of independent reflections to refined parameters.

Caution should be taken while refining some of the parameters simultaneously (i.e. non-targetor non-target, multipole populations and thermal parameters), as they may correlate strongly, resulting in an unstable refinement or unphysical parameter values. Applying additional constraints resulting from local symmetry for each atom in a molecule (which decreases the number of refined multipoles) or importing populational parameters from existing databases may also be necessary to achieve a passable model. On the other hand, the aforementioned approaches significantly reduce the amount of information required from experiments, while preserving some level of detail concerning aspherical charge distribution. Therefore, even macromolecular structures with satisfactory X-ray diffraction data can be modelled aspherically in a similar fashion.

Despite their similarity, individual multipoles do not correspond to any particular molecular orbitals resulting from quantum calculations, as their sole aim is to estimate the charge density distribution, not a molecule's electron structure. In order to find some relationship between the X-ray experiment and quantum mechanics-resulting wave functions, the Hirshfeld Atom Refinement may be applied or other methods of the so-called Quantum Crystallography.

Role in Pyridyl-Ruthenium Complexes 221

A more accurate understanding of the structure leads to the construction of appropriate reaction systems. For example, although the well-known platinum complex, [$PtCl_2(NH_3)_2$], has two geometrical isomers, only the cis-isomer (so-called cisplatin) exhibits prominent antitumor activity.

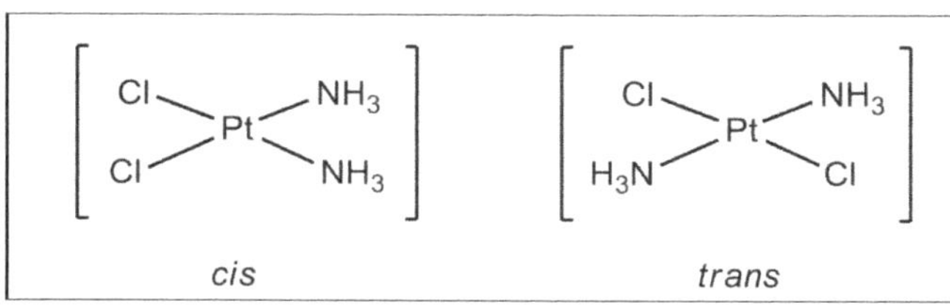

Two geometrical isomers in [$PtCl_2(NH_3)_2$].

Characterization of most structures of simple organic compounds is usually carried out by various spectroscopic measurements including NMR, IR and MS. On the other hand, detailed structural characterization of coordination compounds is achieved by X-ray diffraction methods, which can provide not only information on the geometrical structures, but also full bond parameters. In particular, single crystal X-ray diffraction is the most powerful tool for the detailed structural analysis of crystalline coordination compounds. The aim of this chapter is to demonstrate particular advantages of X-ray structural analysis when compared with other techniques on coordination compounds. Herein, X-ray analysis of a variety of mononuclear ruthenium complexes containing pyridyl substituents is mainly described.

Coordination Geometries and Coordination Modes

Some isomeric pairs of [Ru(tpy)(L)Cl]$^{n+}$ type complexes (tpy = 2,2':6',2"-terpyridine, L = asymmetrical pyridyl-based bidentate ligands) have been prepared and structurally characterized as precatalysts to investigate the effect of isomeric structural features on the catalytic epoxidation process. In the complex [Ru(tpy)(1)(OH_2)]$^{2+}$, the distal isomer exhibits better activity because it contains a pyridine C–H bond nearly parallel to the Ru–O bond, whereas for the proximal isomer this position is occupied by a C–CH_3 group and thus exerts a much stronger steric effect. In the complex [Ru(tpy)(2)(OH_2)]$^+$, on the other hand, the proximal isomer has been established to be an excellent catalyst for the chemoselective epoxidation though limited differences in electronic structural features exist between the isomeric pair. These examples indicate that it is important to clearly distinguish the molecular structures of the compounds. Examples of ruthenium complexes which possess asymmetrical multidentate ligands or multiple donor atoms.

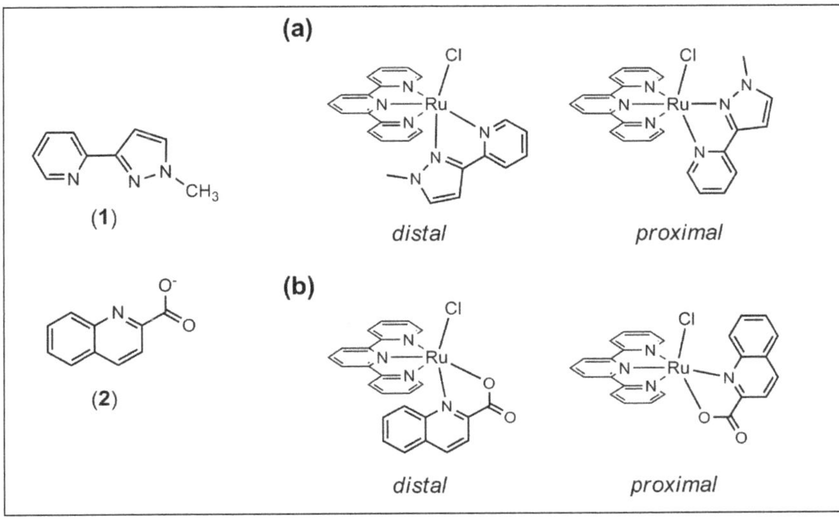

Asymmetrical pyridyl-based ligands (1 and 2) and their ruthenium complexes. (a): $[Ru(tpy)(1)Cl]^+$, (b): [Ru(tpy)(2)Cl].

Coordination Geometries of Azopyridyl Complexes

2-Azopyridyl derivatives behave as mono-, bi- and tridentate ligands, thus a wide variety of mononuclear complexes can be prepared. For example, 2-phenylazopyridine (3), which represents the most fundamental azopyridyl compound, is a bidentate ligand that can coordinate to a metal ion through the lone pairs on the pyridine and the azo nitrogen atoms, thereby forming a stable chelating 5-membered ring.

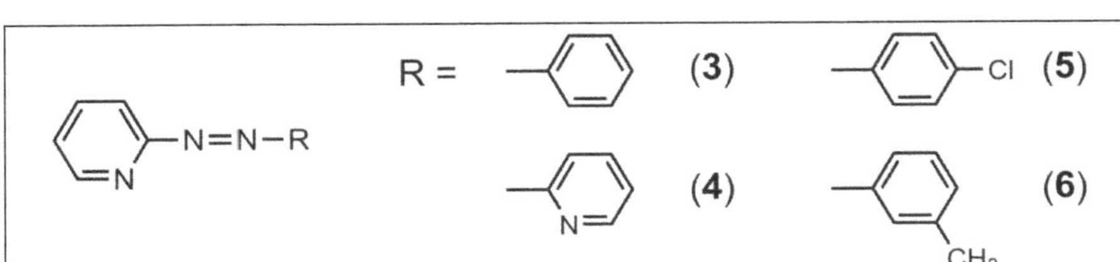

Bidentate 2-azopyridyl ligands.

Since the bidentate ligand (3) lacks a two-fold symmetry axis, there are five possible isomers in $[Ru(3)_2Cl_2]$. Among them, α, β and ε adopt the cis-geometry with respect to the two chlorido ligands, whereas γ and δ are the trans-geometry. Besides the ε isomer, the structures have been determined by X-ray crystallography: the molecular structures of two $[Ru(3)_2Cl_2]$ isomers (α- and β– isomers) were published in 1984, the third (γ) and the fourth (δ) isomers were reported in 2000 and 2004, respectively.

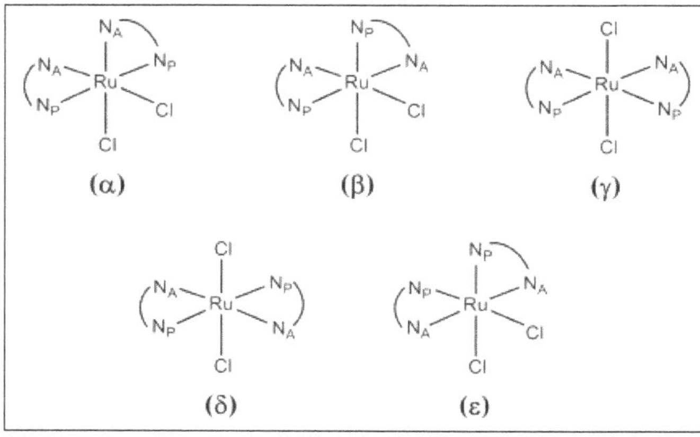

Chemical structures of the five stereoisomers of $[Ru(3)_2Cl_2]$. NA and N_P represent the azo and pyridine nitrogen atoms, respectively.

Ruthenium(II) complexes containing both Ru–C bonds and 3 as supporting ligands were synthesized by substituting chlorido ligand(s) in $[Ru(3)_2Cl_2]$ with CO or CN^-. The molecular structure of $[Ru(3)_2(CO)Cl]^+$ determined by X-ray crystallography is shown in. The geometry of the complex corresponds to the α– isomer shown in figure. The carbonyl C–O and Ru–Cl bond distances, and the Ru–C–O bond angle are consistent with those observed for closely related complexes ($[RuL_2(CO)Cl]^+$: L = bidentate pyridine-based ligands), whereas the Ru–C bond distance is longer than those in similar complexes. This is because of the presence of two 3 ligands with azo moieties. In general, the Ru–N(azo) bond distances are shorter than those of the Ru–N(pyridine) bonds in 3 in the ruthenium complexes. However, one of the Ru–N(azo) bonds is very long; even longer than the distances of the Ru–N(pyridine) bonds. This is caused by the trans influence of the CO ligand.

The crystal structure of $[Ru(3)_2(CN)_2]$ is shown in figure. Unexpectedly, the two CN groups are trans to each other, and the δ–form is observed. A δ arrangement is very rare, and was only recently reported for δ–$[Ru(3)_2Cl_2]$. In the δ–isomer, the two azo bonds (in mutual trans positions) should compete with each other for the Ru(II) 4d electron density, resulting in relatively long Ru–N(azo) bonds in $[Ru(3)_2(CN)_2]$ with respect to the equivalent bonds in γ–$[Ru(3)_2Cl_2]$. This confirms that the Ru–N(azo) bond order decreases in $[Ru(3)_2(CN)_2]$.

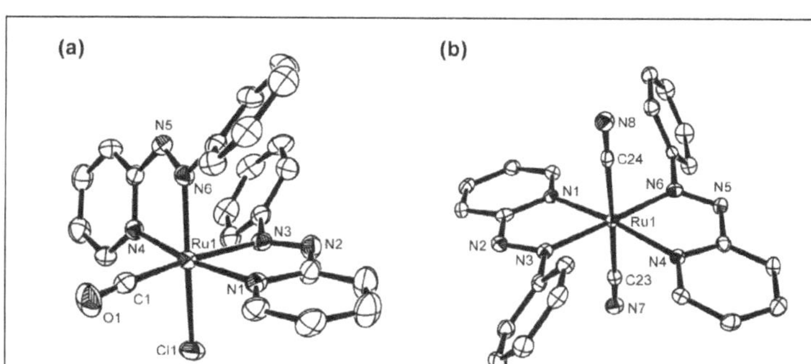

Molecular structures of (a): $[Ru(3)_2(CO)Cl]^+$, (b): $[Ru(3)_2(CN)_2]$.

Although only the *cis*-isomer of $[Ru(3)_2(CO)Cl]^+$ is formed, the *trans*-isomer of $[Ru(3)_2(CN)_2]$ formed preferentially despite the use of a *cis*-$[Ru(3)_2Cl_2]$ precursor. In $[RuL_2(CN)_2]$ (L = bidentate pyridine-based ligands) complexes, the *cis* form is the preferred geometry for 2,2'- bipyridine and its analogues, whereas the *trans* geometry is more stable in $[Ru(3)_2(CN)_2]$.

Many analogous bis-azopyridine complexes, $[Ru(azopyridine)_2L_2]^{n+}$, have also been reported. In these complexes, the $[Ru(4)_2(CO)(PPh_3)]^{+/o}$ complexes form the unique ε–isomer examples.

In analogy with the bis-azopyridine complexes, the geometry of the complexes should also be consistent for the mono-azopyridine complexes because of the asymmetrical units of the azopyridines. For example, ruthenium complexes containing the ligand 3 and its analogue 5 ($[Ru(azopyridine)(H)(CO)(PPh_3)_2]$) exhibit two different geometries. Therefore, it is important to control the orientation of the azopyridine ligands in preparing complexes that rely on the steric and/or electronic nature of the ligands. Such examples have been demonstrated in $[Ru(3)(bpy)Cl(L)]^+$ (bpy = 2,2'-bipyridine, L = CO or CH_3CN) complexes. Here, the configuration of the incoming azopyridine ligand 3 can be regulated by the strength of the π–bonding ability of the co-ligands L situated in the *trans* position: the pyridine nitrogen (weaker π–bonding character) of the incoming ligand 3 is selectively situated *trans* to the L ligand when L is a carbonyl, whereas the azo nitrogen (stronger π–bonding character) is situated *trans* to the acetonitrile ligand.

Thus, the combination of CO–pyridine nitrogen (strong-weak π–bonding character) and CH_3CN–azo nitrogen(weak-strong π–bonding character) are quite reasonable. The molecular structures of $[Ru(3)(bpy)Cl(L)]^+$ (L = CO or CH_3CN) are shown in figure. The azo group of 3 in $[Ru(3)(bpy)Cl(L)]^+$ is clearly situated *cis* (L = CO) and *trans* (L =CH_3CN) to the L ligand. The corresponding complex with 4 ([Ru(4)(bpy)(CO)Cl]+) is also selectively synthesized by using the above methodology.

Molecular structures of various ruthenium mono-terpyridine complexes of the type $[Ru(tpy)(azopyridine)L]^{n+}$ (azopyridine = 3, 4, 6; L = Cl, OH_2, CH_3CN, NO_2) have been determined by X-ray analysis. The asymmetrical nature of azopyridines lead to the possibility of two isomers as, however, only one isomer has been characterized in all complexes.

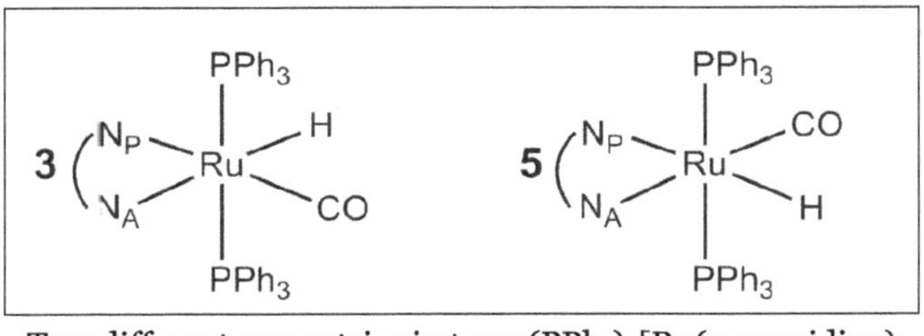

Two different geometries in trans(PPh_3)-[Ru(azopyridine)(H)(CO)$(PPh_3)_2$] (azopyridine = (3) or (5)).

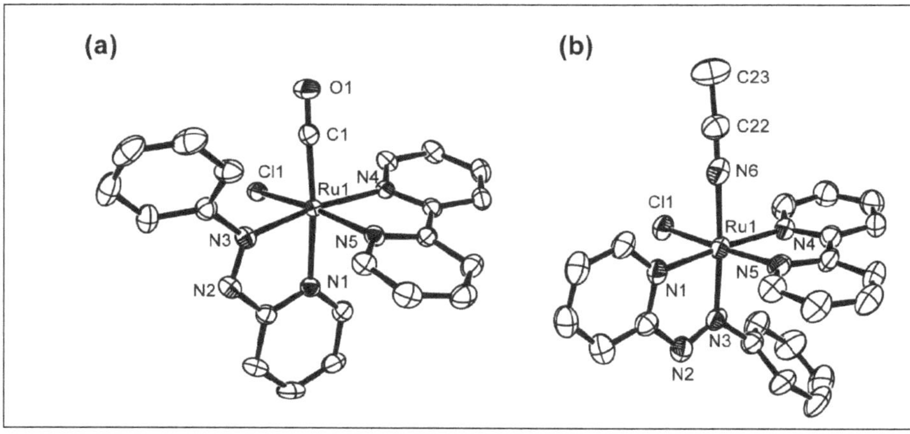

Molecular structures of (a): $[Ru(3)(bpy)(CO)Cl]^+$, (b): $[Ru(3)(bpy)(CH_3CN)Cl]^+$.

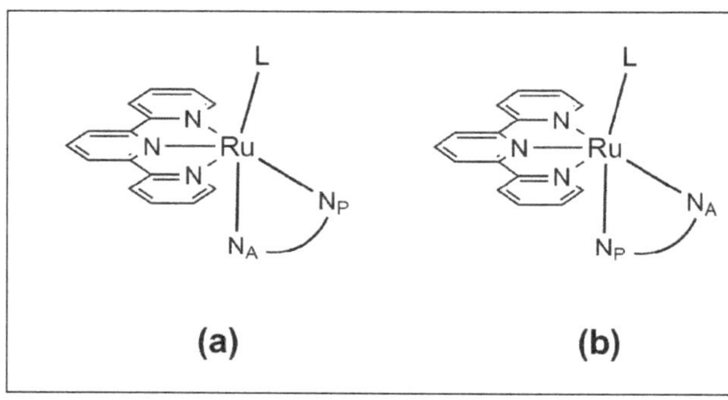

Possible isomers in [Ru(tpy)(azopyridine)L] n^+ type complexes.

Finally, tris-bidentate and bis-tridentate azopyridine complexes are presented. The compound $[Ru(azopyridine)_3]$ n^+ theoretically exists in two different isomeric forms (disregarding Δ, Λ enantiomerism), i.e., the meridional (mer) and facial (fac) isomers. Only mer-$[Ru(azopyridine)_3]^{2+}$ were obtained and crystallographically characterized due to the facial isomer of $[Ru(azopyridine)_3]^{2+}$ being sterically unfavorable because of the crowding of the three phenyl groups. Ruthenium complexes with tridentate azopyridine ligands in figure. have been reported. The configuration is bis-meridional, and the relative orientations of the pairs of coordinated atoms are cis (pyridyl–N), trans (azo–N) and cis (arylanilido–N), respectively.

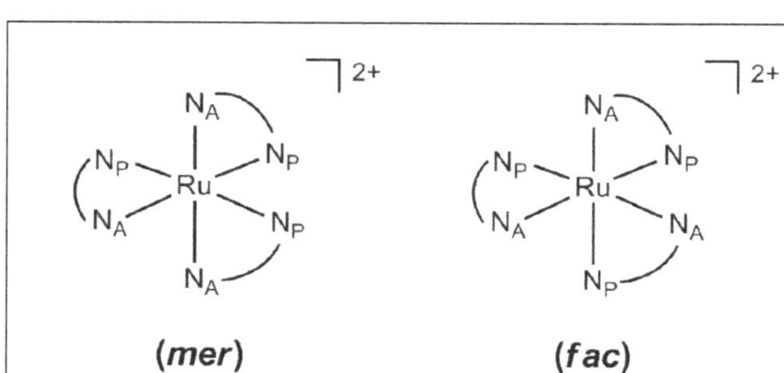

Possible isomers in $[Ru(azopyridine)_3]^{n+}$ type complexes.

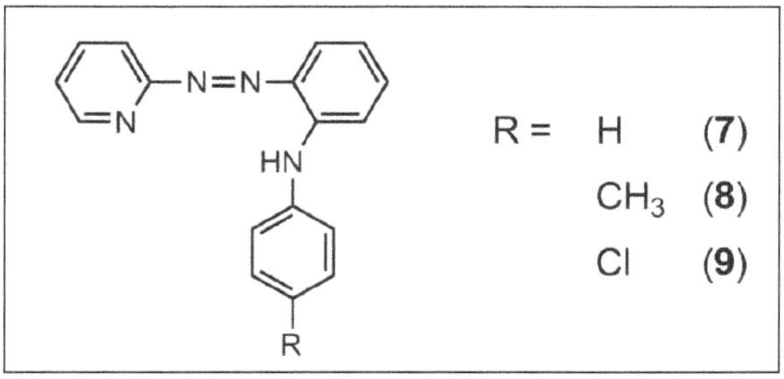

Tridentate 2-azopyridyl ligands.

Coordination Geometries of other Pyridyl-based Complexes

The naphthyridines consist of a group of diazanaphthalenes with one nitrogen in each ring. In particular, the bidentate naphthyridine, 2-(2-pyridyl)-1,8-naphthyridine (pynp; 10), is a useful ligand for mononuclear systems.

For complexes with ligand 10; however, there is a stereochemical question with regard to the binding of ligand 10 because it has an asymmetrical structure. Consequently, it would be very interesting to study complexes involving 10 in an effort to understand the relationship between the coordination geometries of these complexes and their reactivities.

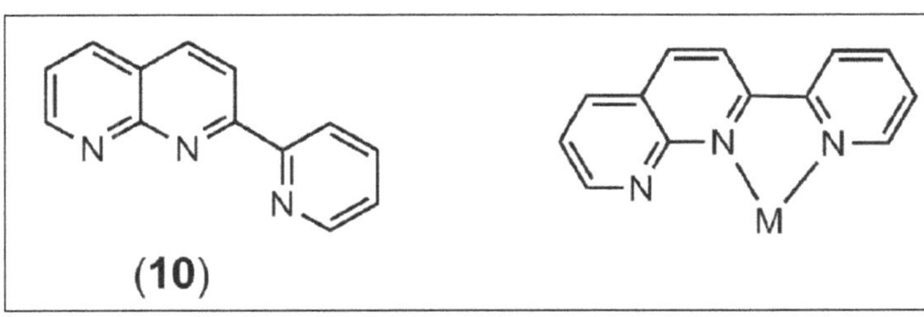

Chemical structure of 2-(2-pyridyl)-1,8-naphthyridine (pynp) and its coordination to a metal center.

The $[Ru(tpy)(10)Cl]^+$ complex acts as a catalyst for water oxidation: it shows excellent catalytic activity with turnover numbers (TNs) of 1,170. This value is the highest in the analogous $[Ru(tpy)(L)Cl]^+$ type complexes (L = bidentate polypyridyl ligands) whose TNs are in the range of 0–570. The ligand 10 differs from all of the other bidentate (L) because it is asymmetric.

Although there are two possible isomers in $[Ru(tpy)(10)Cl]^+$, only one isomer is isolated and its structure has been determined by X-ray crystallography. Two isomeric mixtures of $[Ru(tpy)(10)Cl]^+$ exist in solution (approximately 1:1), both of which can be separated by column chromatography. Alternatively, irradiation of visible light to isomeric mixtures of $[Ru(tpy)(10)L]^{2+}$ (L = OH_2 or CH_3CN) leads the formation of only one isomer in particular organic solvents. The resulting complex ($[Ru(tpy)(10)(OH_2)]^{2+}$) is the proximal configuration, which differs from the corresponding chlorido complex.

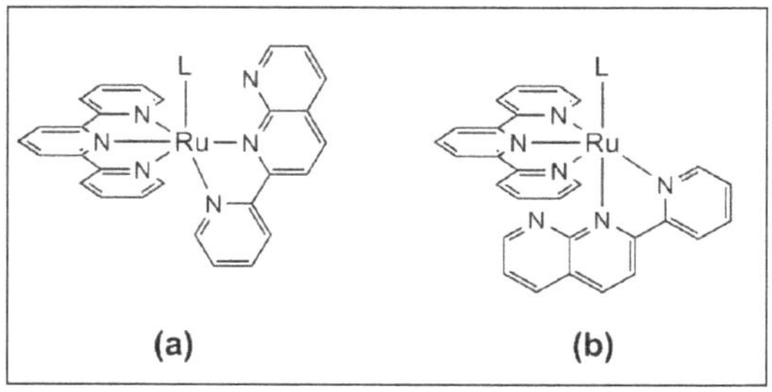

Possible isomers in $[Ru(tpy)(10)L]^{n+}$ type complexes.

In contrast, particular stereoisomers could be selectively prepared in the cis- [Ru(bpy)(10)(CO)Cl]$^{+}$ system. As shown in figure, although there are four possible geometries in the complex, only two types of the complex could be selectively synthesized. Although the prepared complexes are all single species as determined from the spectroscopic measurements, their structures cannot be assigned because of their spectral resemblance. Consequently, X-ray analysis was required to determine their detailed structures. The molecular structures of two [Ru(bpy)(10)(CO)Cl]$^{+}$ isomers are shown in figure. One isomer corresponds to (a) in figure. (the pyridyl ring of 10 is situated trans to the CO ligand), the another isomer corresponds to (b) in figure. (the pyridyl ring of 10 is situated cis to the CO ligand).

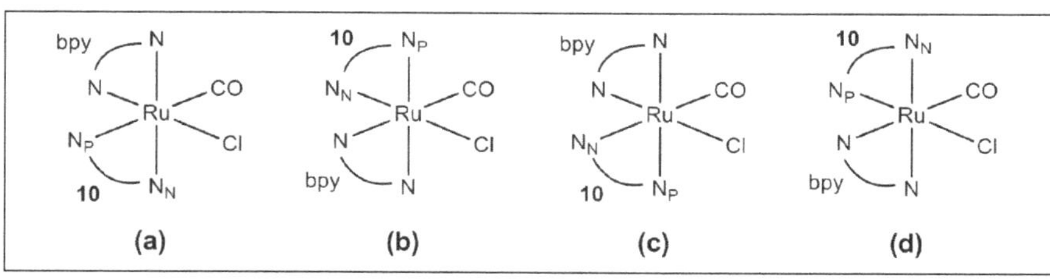

Chemical structures of four stereoisomers in cis-[Ru(bpy)(10)(CO)Cl]$^{+}$. NN and NP denote the naphthyridine nitrogen and the pyridine nitrogen, respectively.

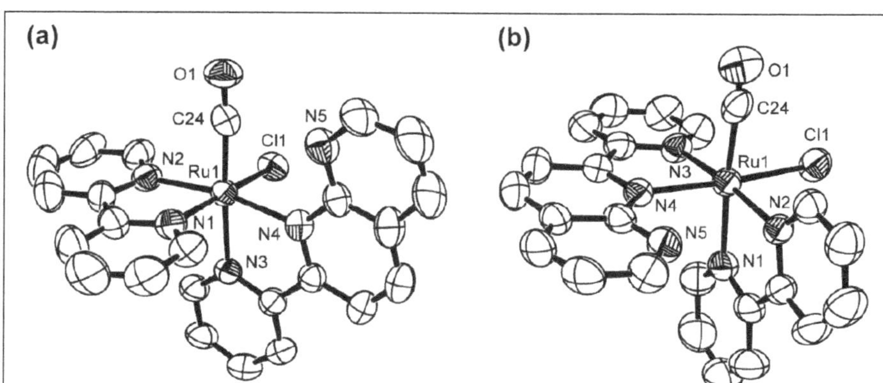

Molecular structures of [Ru(bpy)(10)(CO)Cl]$^{+}$. (a): isomer "a".: isomer "b".

Picolinato can also coordinate to a metal ion as an asymmetrical bidentate ligand. The mono-pic complex, $[Ru(11)(CO)_2Cl_2]^-$, has been prepared and fully characterized, including X-ray analysis. Although three geometrical isomers are expected for [Ru(11)$(CO)_2Cl_2]^-$, only form (i) in figure was confirmed, and neither form (ii) nor (iii) was detected in the solid state and in solutions. The asymmetrical chelate of 11 gives substantially different effects to the two Ru–CO bonds of $[Ru(11)(CO)_2Cl_2]^-$. The Ru–C bond distance trans to the oxygen atom is shorter than another Ru–C bond trans to the nitrogen atom. The complex containing mono-11, cis- $[Ru(bpy)(11)(CO)_2]^+$ was also prepared by the reaction of Hpic with $[Ru(bpy)(CO)_2Cl_2]$, and only the type (iv) of the two possible isomers (iv) and (v) was obtained.

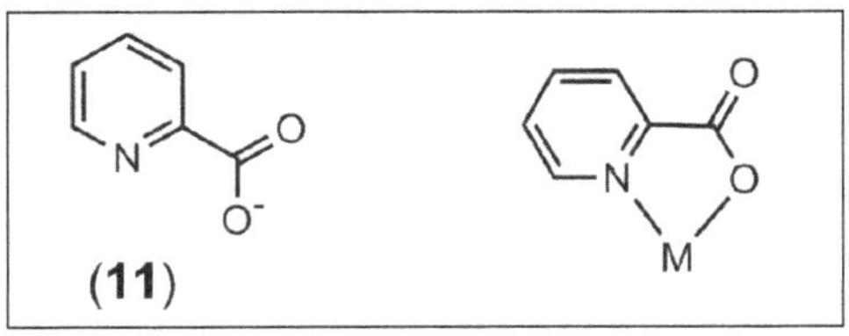

Chemical structure of picolinato (pic) and its coordination to a metal center.

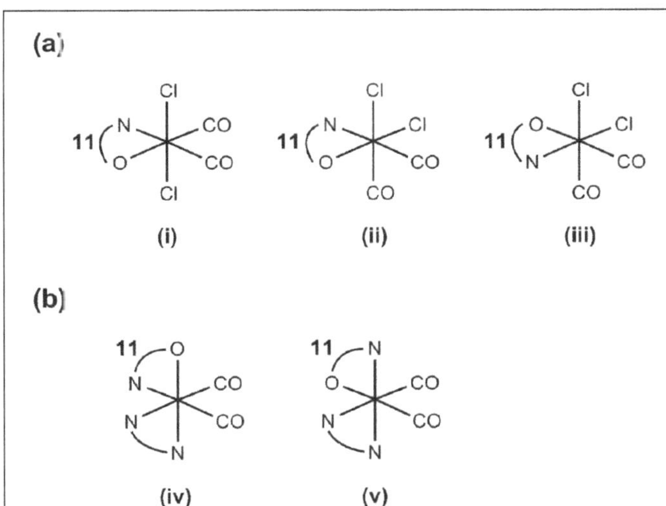

Possible geometrical isomers containing 11. (a): $[Ru(11)(CO)_2Cl_2]^-$, (b): cis- $[Ru(bpy)(11)(CO)_2]^+$.

Two geometrical isomers of $[Ru(tpy)(bpyO)(CO)]^+$ (bpyO = 2,2'-bipyridin-6-onato; 12) were selectively synthesized by different synthetic routes. These isomers bearing 12 were determined by X-ray crystal structure analysis. Based on their pKa's, redox potentials and IR spectra, the electron density at the ruthenium center of the distal isomer in figure. is higher than that of the proximal isomer. Such a difference is associated with the stronger trans influence of CO compared with the central pyridine of tpy, because the electron-withdrawing ability of CO induces the pyridonato structure of the resonance between ruthenium-pyridionate (Ru–bpyO–) and –pyridonato (Ru–bpyO) forms. The two-electron reductions of the distal isomer are followed by a partial Ru–CO bond cleavage, whereas the 12-based reduction of the proximal isomer causes cyclometallation by an attack of the pyridonato oxygen at the carbonyl carbon.

Thus, cyclometallation caused by 12-based reduction effectively suppresses the reductive cleavage of a Ru–CO bond.

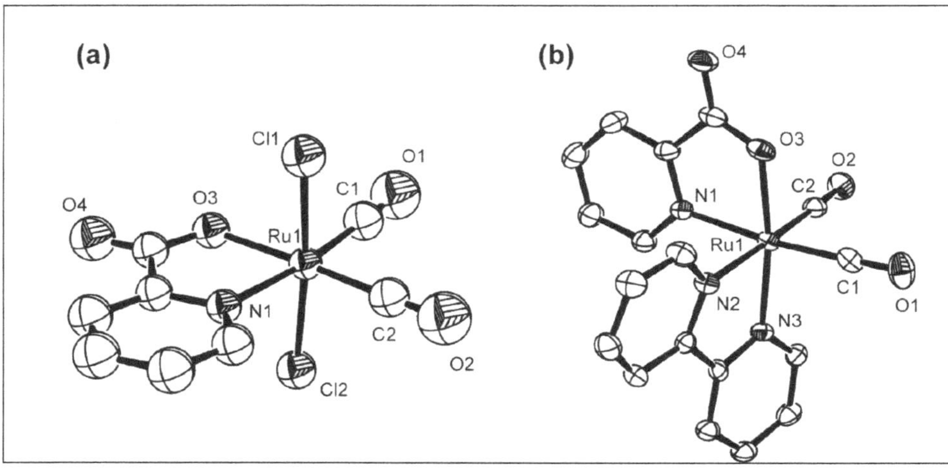

Molecular structures of (a): $[Ru(11)(CO)_2Cl_2]^-$ (isomer (i), (b): $[Ru(bpy)(11)(CO)2]^+$ (isomer (iv).

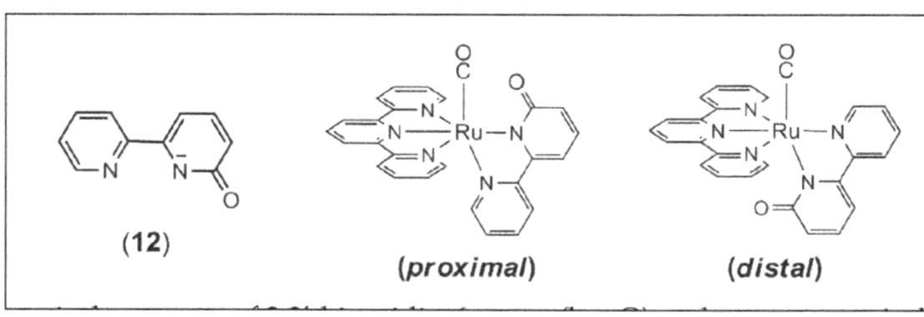

Chemical structure of 2,2'-bipyridin-6-onato (bpyO) and two geometrical isomers in $[Ru(tpy)(12)(CO)]^+$.

Coordination Modes

It is known that not only some acido anions such as NO_3^-, CO_2R^- and CO_3^{2-} but also organic ligands having multi donor atoms, can coordinate to metals as monodentates, chelating bidentates, or bridging bidentate ligands. In coordination compounds, the NO_3^- ion can be noncoordinating (counter ion), or it can coordinate in a monodentate (NO_3^-O) or bidentate (NO_3^-O,O') fashion. The N–O stretching vibrations, for example, are observed at 1475, 1272 and 991 cm–1 in the IR spectrum of cis- $[Ru(bpy)_2(CO)(NO_3)]$PF6. It is rather difficult to differentiate between these structures by vibrational spectroscopy because the symmetry of the NO_3– ion differs very little between them (C_s vs. C_{2v} symmetry). However, X-ray data clearly indicates the monodentate coordination of the NO_3^- ion. The sums of the three O–N–O bond angles in the nitrato ligand are 360.0°, indicating that the nitrato moiety has a planar structure.

Extensive studies have been made on metal complexes of carboxylic acids. The carboxylate ion may also coordinate to a metal in a monodentate (CO_2R–O) or bidentate

(CO_2R–O,O') mode. In general, the coordination mode of the carboxylate ion is distinguishable using vibrational spectroscopy. Monodentate complexes exhibit Δν values ($\nu_{as}(CO_2^-)-\nu_s(CO_2^-)$) which are much greater than the ionic compounds, whereas chelating (bidentate) complexes exhibit Δν values which are significantly less than the ionic values. For example, the cis-[Ru(bpy)2(CO)(CO_2H)]$^+$ complex shows the large formato Δν value of 339 cm^{-1}, and this value is clearly greater than the ionic value (Δν = 201 cm^{-1}). X-ray structural data for this complex clearly identified the monodentate formato ligand (CO_2H–O) bound to an octahedral ruthenium center.

The azopyridine ligand (3) generally coordinates to a metal ion through both pyridyl and azo nitrogens (bidentate chelate fashion). However, the crystal structure of a ruthenium complex containing 3 as a monodentate ligand ([Ru(bpy)$_2$(3–N)(CO)]$^{2+}$) has been reported. The ligand 3 coordinates to the ruthenium center through only the pyridyl nitrogen atom. The remarkable feature in the structure of this complex concerns the azo moiety of 3, which is directed toward the adjacent terminal carbonyl ligand. Taking into account the fact that the nitrogen atom of the azo group is located just to the side of the carbonyl carbon, the existence of the CO ligand could influence the orientation of the azo moiety of 3. The redox reactions of [Ru(bpy)2($_3$– N)(CO)] $^{2+}$ using solution IR spectra under electrolysis conditions revealed a nucleophilic attack of the azo nitrogen atom of 3 on the carbonyl carbon, because of the short interatomic distance between the azo nitrogen and the carbonyl carbon atoms (2.865 Å). In contrast, the dissociation of CO from the two-electron reduced species easily occurs in the analogue with the bidentate 3, because the ligand is unable to interact with CO.

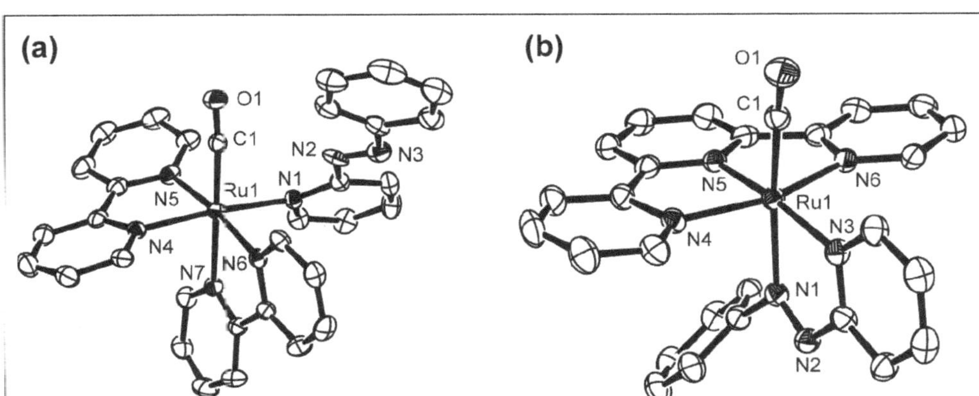

Molecular structures of (a): [Ru(bpy)2(3–N)(CO)]$^{2+}$, (b): [Ru(tpy)(3–N,N')(CO)]$^{2+}$.

The diphenyl-2-phosphinopyridine ligand (dppy; 13) has been used to construct interesting metal complexes because it possesses both the soft (P) and hard (N) donor atoms. Although both the P-mono and P,N-bidentate coordination modes have been known for mononuclear systems, 13 coordinates to the ruthenium center through only the P atom in [Ru(bpy)(13)(CO)$_2$Cl]$^+$. A prominent feature of the structure is that the pyridyl nitrogen atom of 13 directs toward the plane that includes two carbonyl ligands. The existence of carbonyl ligands may influence the orientation of the aromatic rings of

13 because relatively short distances between the noncoordinating nitrogen atom of 13 and the carbonyl carbons are observed (3.068 Å): the analogous complex situating two 13 moieties in mutually trans positions also exhibits a similar trend.)

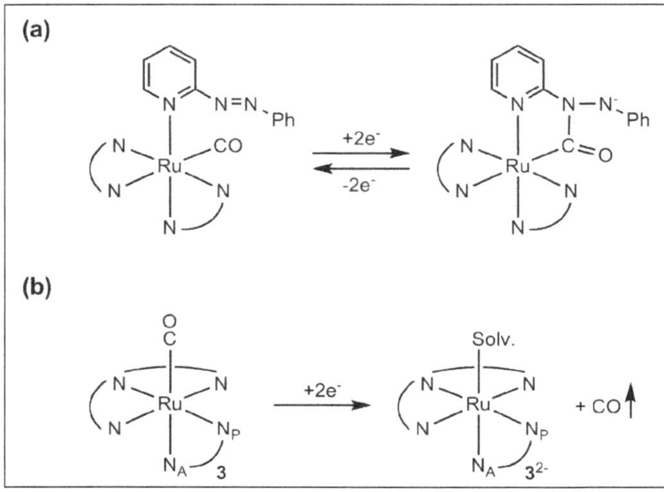

The different redox behavior of the ruthenium complexes are dependent on the coordination mode of 3. (a): $[Ru(bpy)2(3–N)(CO)]^{2+}$ (monodentate 3), (b): $[Ru(tpy)(3– N,N')(CO)]^{2+}$ (bidentate 3).

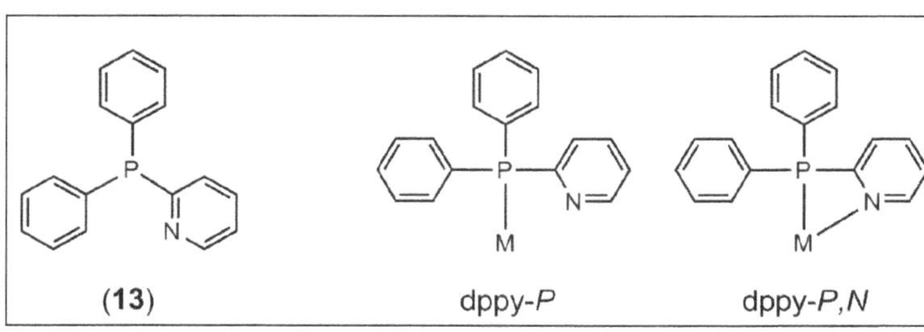

Two coordination modes of diphenyl-2-phosphinopyridine (dppy).

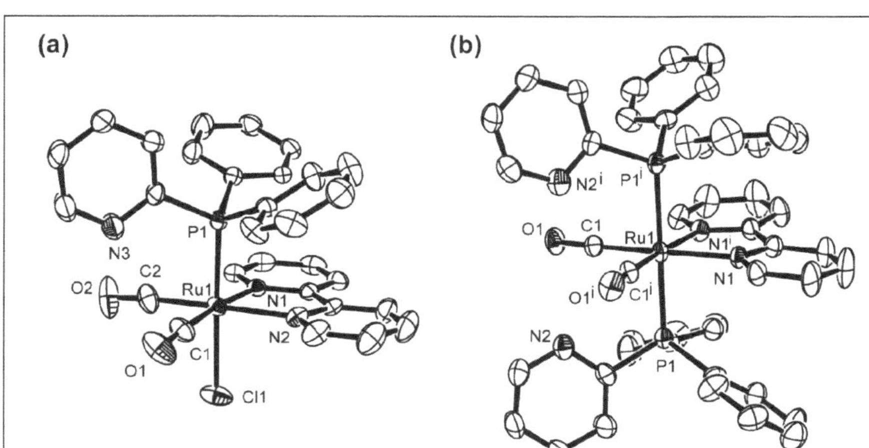

Molecular structures of (a): $[Ru(bpy)(13–P)(CO)_2Cl]^{+}$,
(b): $([Ru(bpy)(13–P)2(CO)_2]^{2+}$.

Comparison of the Structures and Structural Parameters

When chemists simply recognize the molecular structure of a compound, they usually pay particular attention to the points described above. However, it is also very helpful for chemists to understand how intermolecular forces (packing forces) act between molecules in crystals when they investigate the constructions of materials and medicines. In addition, various chemical and physical properties can be predicted by comparing bond parameters (e.g., bond distances, angles or torsion angles) in a compound.

Comparison of Structures

Almost all structurally characterized coordination compounds of azopyridines have the ligand bonded to one metal center via the azoimine (N=N–C=N) chelate arrangement. In the case of 3 and 4, formation of a chelate complex creates the free aromatic ring (phenyl for 3 or 2-pyridyl for 4). The orientation of the pendant aromatic ring is dependent on its coordination environment: the dihedral angle between the two aromatic rings of $[Ru(3)(CO)_2Cl_2]$ is 35.29°) whereas the dihedral angle between the two pyridyl rings in $[Ru(4)(CO)_2Cl_2]$ is 7.99°), which shows planarity of the coordinated 4. The distinction observed in solid state can be explained on the basis of a possible weak interaction between the non-coordinating nitrogen atom and the adjacent carbonyl carbon in $[Ru(4)(CO)_2Cl_2]$ rather than a different bulkiness between the CH in phenyl ring of 3 and the N in pyridyl ring of 4, because the interatomic distance between the N atom and the adjacent C atom in $[Ru(4)(CO)_2Cl_2]$ is fairly short (2.682(3) Å). This weak interaction has often been observed in ruthenium carbonyl complexes with noncoordinating pyridyl moieties, which contain both a pyridyl nitrogen (δ-) and a carbonyl carbon (δ+).

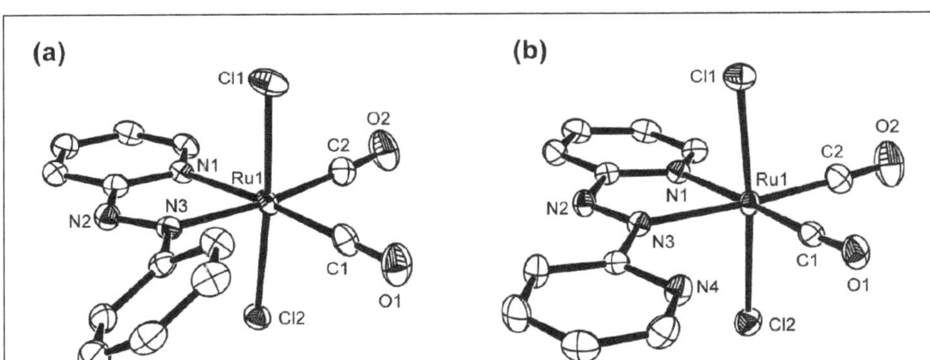

Molecular structures of $[Ru(azopyridine)(CO)_2Cl_2]$. (a): $[Ru(3)(CO)_2Cl_2]$, (b): $[Ru(4)(CO)_2Cl_2]$.

The X-ray structures of terpyridyl complexes with the dimethoxyphenyl pendant have been determined. In the metal-free ligand 14, the three pyridyl rings are approximately coplanar, whereas the dimethoxyphenyl substituent in 14 is not coplanar with the terpyridyl moiety, making an angle of 50.2° with the central pyridyl ring. The dihedral angle between the central pyridyl ring and the dimethoxyphenyl ring in the ruthenium complexes with 14 are in the range of 43 to 56°, which are close to that of metal-free ligand 14. Interestingly, the OCH_3 orientation of two methoxy moieties are different in two types of habits (yellow needles and red blocks) of $[Ru(14)(bpy)(CO)](PF_6)_2$. As shown in figure, two OCH_3 moieties point in the same orientation in the yellow crystal, whereas in the red crystal they point in opposite directions. The former structure is identical to the corresponding chlorido and acetonitrile complexes. On the other hand, the latter form is consistent with that observed in the metal-free. These differences are most likely due to different packing or intermolecular effects on the orientation of methoxy groups of 14.

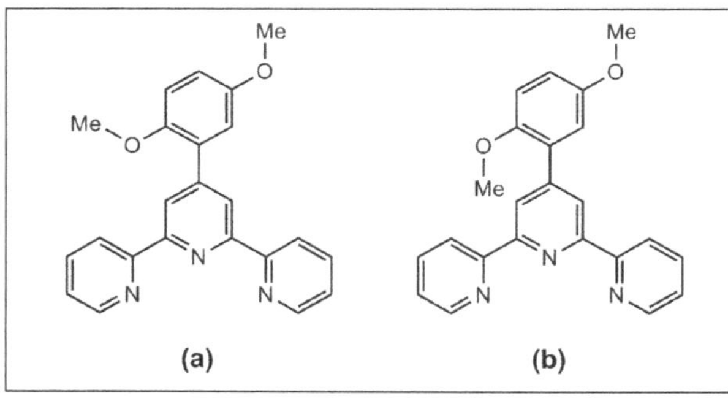

Chemical structures of 14: (a): same orientation, (b): opposite orientation, regarding two methoxy groups.

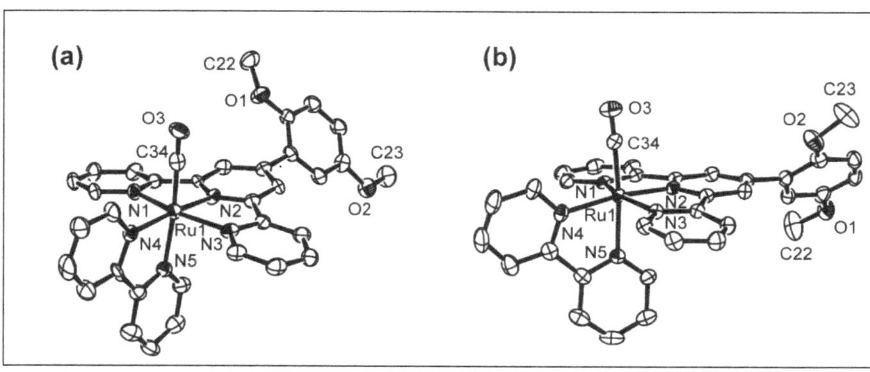

Molecular structures of $[Ru(14)(bpy)(CO)]^{2+}$: (a): the yellow crystal, (b): the red crystal.

Comparison of Structural Parameters

Physical data of terminal carbonyl groups are generally useful indicators for electronic states around metal centers in metal complexes. In particular, bond distances and vibrational spectra are considered important. In order to distinguish the electron states in the ruthenium mono carbonyl complexes with polypyridyl ligands, the stretching force constants of the CO group using IR spectral data were calculated in addition to bond distances and stretching frequencies of the $\{Ru–CO\}^{2+}$ moiety. Although obvious characteristics about the bond distances (Ru–C and C–O) are not shown in the complexes, the force constant (k) and the stretching frequencies of the CO groups can clearly be divided into two groups: the k values of the azocontaining complexes are higher ca. 0.4–1.1 N cm–1 than of the values for the complexes without azopyridines. It is presumed that the electron density of the ruthenium center considerably decreases when an azo group (which has greater π–acidity) coordinates to the ruthenium atom, because the decrease of the k value caused by ligand-based one-electron reduction corresponds to 0.56 N cm^{-1} in $[Ru(bpy)_2(CO)(quinoline)]^{2+}$. Thus, it can be concluded that complex formation with azopyridyl ligands induces relatively high k values. This observation can be ascribed to the decrease of the π backdonation from the ruthenium center to the CO group. As a result, this suggests that the azopyridyl ligands serve as prominent electron reservoirs compared with other polypyridyl ligands such as bpy.

Table: Bond distances (d: Å), stretching frequencies (v: cm–1) and force constant (k: N cm^{-1}) of the CO groups in cis-$[Ru(polypyridine)_2(CO)Cl]^+$.

Polypyridines	d(Ru–C)	d(C–O)	vCO	k
3/3	1.897(5)	1.136(6)	2031	16.67
4/bpy	1.889(3)	1.116(4)	2002	16.20
3/bpy	1.891(2)	1.122(3)	2001	16.18
4,4'-Me_2bpy/5,6'-Me_2phen	1.867(5)	1.112(7)	1978	15.81
dpk/4,7'-Me_2phen	1.820(11)	1.146(13)	1976	15.80
10/bpy	1.867(5)	1.145(6)	1975	15.76
dpa/5,5'-Me_2bpy	1.917(4)	1.026(6)	1972	15.72
bpy/dpk	1.880(5)	1.119(7)	1970	15.68
4,4'-Me_2bpy/4,7'-Me_2phen	1.95(2)	1.05(2)	1967	15.64
bpy/dpa	1.883(16)	1.08(2)	1967	15.64
bpy/phen	1.857(4)	1.126(6)	1966	15.62
bpy/bpy	1.861(29)	1.122(32)	1966	15.62
biq/biq	1.91(2)	1.01(2)	1964	15.59

The N–N bond distance in an azo moiety is an excellent indicator of the charge on an azo group. The values for unreduced azo N–N bonds are 1.22–1.31 Å in metal complexes (1.23–1.26 Å for metal-free ligands). The one-electron reduced (anion radical) ligands have a bond distances of 1.31–1.41 Å, whereas the two-electron reduced (hydrazido) forms have single bonds with 1.41–1.50 Å distances. As shown in Table, the azo N–N bond distance of [Ru(bpy)2(CO)($_3$–N)]$^{2+}$ is dramatically shorter than those of other complexes: the distance (1.188(4) Å) is shorter than that of a typical N=N double bond (1.23 Å).

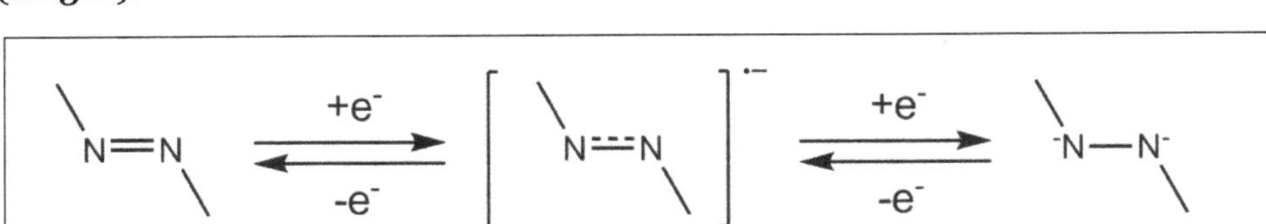

Three oxidation states of an azo moiety.

Table: Bond distances (å) in the complexes of mono-azopyridines and metal-free ligands.

Compound	d(N–N)	d(M–N_{azo})	d(M–N_{py})
[Ru(3)$(CO)_2Cl_2$]	1.264(3)	2.119(2)	2.099(2)
[Ru(4)$(CO)_2Cl_2$]	1.265(3)	2.107(2)	2.108(2)
[Ru(3)(bpy)(CO)Cl]PF_6	1.287(3)	2.015(2)	2.095(2)
[Ru(4)(bpy)(CO)Cl]PF_6	1.291(4)	1.998(2)	2.114(2)
[Ru(3)(tpy)(CO)]$(ClO_4)_2$	1.265(5)	2.099(3)	2.071(3)
[Ru(4)(tpy)Cl]PF_6	1.299(3)	1.964(2)	2.054(2)
[Ru(3– κ N)$(bpy)_2$(CO)]$(ClO_4)_2$	1.188(4)		2.131(2)
($3H^+$)BF_4	1.248(4)		
($3H^+$)ClO_4	1.258(5)		
4	1.246(2)		

In the case of bis-azopyridyl complexes, bivalent azopyridine properties are observed by the different azo N–N bond distances. For example, the structure of the monoradical [Ru(4)(4i •–)(CO)(PPh_3)]$^+$ has been determined. The most remarkable feature of the structure is the difference in the two N–N bond distances: one is 1.284(6) Å, the other N–N bond is 1.336(6) Å. Accordingly, both unreduced and one-electron reduced (anion radical) types are present within the same molecule, which represents a model case of ground-state radical localization in one of the two ligands. The stabilization of radical on one ligand might be supported by the trans ligand. A similar example is also reported in [Ru$(7^{2-})_2$]. The [Ru$(7)_2$] molecule has longer N–N bond distances (1.324(7) and 1.327(6) Å), clearly suggesting the radical dianion form of the ligand (one charge from the deprotonation and another charge from the one-electron reduction centered on the azo group). This diamagnetic molecule may thus be viewed as a singlet diradical species. In this situation, the electrons from a d^4 configuration at the metal couple with the singly occupied ligand molecular orbitals to create a spin-paired entity. This situation has been confirmed by DFT (density functional theory) calculations.

Cryo Bio-crystallography

Cryo bio-crystallography is the application of crystallography to biological macromolecules at cryogenic temperatures.

Basic Principles

Cryo crystallography enables X-ray data collection at cryogenic tempertures, typically 100K:

- Crystals are transferred from the solution they have grown in (called mother liquor) to a solution with a cryo-protectant to prevent ice formation.

- Crystals are mounted with a glass fiber (as opposed to a capillary).
- Crystals are cooled by dipping directly into liquid nitrogen and then placed in a cryo cold stream.
- Cryo cooled macromolecular crystals show reduced radiation damage by more than 70 times that at room temperature.

Advantages

- Significant improvement of resolution in data collection.
- Reduced or eliminated radiation damage in crystals.

Usefulness and Applications

Crystallography of large biological macromolecules can be achieved while maintaining their solution state. The most known example is the ribosome.

Optical Crystallography

Optical crystallography is the branch of crystallography that deals with the optical properties of crystals. It is of considerable interest theoretically and has the greatest practical importance. The science of petrography is largely based on the study of the appearance of thin, transparent sections of rocks in a microscope fitted with polarizers. In the absence of external crystalline form, as with the minerals in a rock, a mineral often may be readily identified by the determination of some of its optical properties.

According to their action on transmitted plane-polarized light, all crystals may be assigned to one or another of the five groups enumerated below, which correspond to the seven systems of crystallization (in the second group, three systems are included together). The several symmetry classes of each system are optically the same, except in rare cases:

- Optically isotropic crystals (isometric system), which exhibit only one index of refraction for light of each colour.
- Optically uniaxial crystals (tetragonal, hexagonal, and rhombohedral [or trigonal] systems), which exhibit double refraction and yield two refractive indices for light of each colour, one parallel to the optical axis and one perpendicular to the optical axis.
- Optically biaxial crystals (all of which exhibit three principal refractive indices, one along each of the mutually perpendicular optical axes) in which the three

optical axes correspond to the three crystallographic axes (orthorhombic system).

- Optically biaxial crystals in which only one of the three optical axes corresponds to a crystallographic axis (monoclinic system).
- Optically biaxial crystals with no fixed and definite relation between the optical and crystallographic axes (triclinic system).

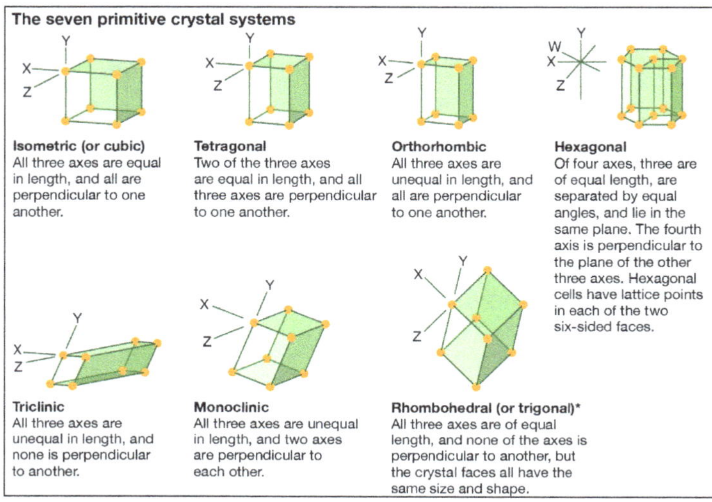

Crystal systems: The seven primitive crystal systems.

Electron Crystallography

Electron crystallography is a method to determine the arrangement of atoms in solids using a transmission electron microscope (TEM).

Comparison with X-ray Crystallography

It can complement X-ray crystallography for studies of very small crystals (<0.1 micrometers), both inorganic, organic, and proteins, such as membrane proteins, that cannot easily form the large 3-dimensional crystals required for that process. Protein structures are usually determined from either 2-dimensional crystals (sheets or helices), polyhedrons such as viral capsids, or dispersed individual proteins. Electrons can be used in these situations, whereas X-rays cannot, because electrons interact more strongly with atoms than X-rays do. Thus, X-rays will travel through a thin 2-dimensional crystal without diffracting significantly, whereas electrons can be used to form an image. Conversely, the strong interaction between electrons and protons makes thick (e.g. 3-dimensional > 1 micrometer) crystals impervious to electrons, which only penetrate short distances.

One of the main difficulties in X-ray crystallography is determining phases in the diffraction pattern. Because of the complexity of X-ray lenses, it is difficult to form an image of the crystal being diffracted, and hence phase information is lost. Fortunately, electron microscopes can resolve atomic structure in real space and the crystallographic structure factor phase information can be experimentally determined from an image's Fourier transform. The Fourier transform of an atomic resolution image is similar, but different, to a diffraction pattern—with reciprocal lattice spots reflecting the symmetry and spacing of a crystal. Aaron Klug was the first to realize that the phase information could be read out directly from the Fourier transform of an electron microscopy image that had been scanned into a computer, already in 1968. For this, and his studies on virus structures and transfer-RNA, Klug received the Nobel Prize for chemistry in 1982.

Radiation Damage

A common problem to X-ray crystallography and electron crystallography is radiation damage, by which especially organic molecules and proteins are damaged as they are being imaged, limiting the resolution that can be obtained. This is especially troublesome in the setting of electron crystallography, where that radiation damage is focused on far fewer atoms. One technique used to limit radiation damage is electron cryomicroscopy, in which the samples undergo cryofixation and imaging takes place at liquid nitrogen or even liquid helium temperatures. Because of this problem, X-ray crystallography has been much more successful in determining the structure of proteins that are especially vulnerable to radiation damage. Radiation damage was recently investigated using MicroED of thin 3D crystals in a frozen hydrated state.

Application to Inorganic Materials

Electron diffraction pattern of the same crystal of inorganic tantalum oxide shown above. Notice that there are many more diffraction spots here than in the diffractogram calculated from the EM image above. The diffraction extends to 12 orders along the 15 Å direction and 20 orders in the perpendicular direction. Thus the resolution of the ED pattern is 1.25 Å (15/12 or 25/20). ED patterns do not contain phase information, but the clear differences between intensities of the diffraction spots can be used in crystal structure determination.

Electron crystallographic studies on inorganic crystals using high-resolution electron microscopy (HREM) images were first performed by Aaron Klug in 1978 and by Sven Hovmöller and coworkers in 1984. HREM images were used because they allow to select (by computer software) only the very thin regions close to the edge of the crystal for structure analysis. This is of crucial importance since in the thicker parts of the crystal the exit-wave function (which carries the information about the intensity and position of the projected atom columns) is no longer linearly related to the projected crystal structure. Moreover, not only do the HREM images change their appearance with increasing crystal thickness, they are also very sensitive to the chosen setting of the defocus Δf of the objective lens (see the HREM images of GaN for example). To cope with this complexity Michael O'Keefe started in the early 1970s to develop image simulation software which allowed to understand an interpret the observed contrast changes in HREM images.

There was a serious disagreement in the field of electron microscopy of inorganic compounds; while some have claimed that "the phase information is present in EM images" others have the opposite view that "the phase information is lost in EM images". The reason for these opposite views is that the word "phase" has been used with different meanings in the two communities of physicists and crystallographers. The physicists are more concerned about the "electron wave phase" - the phase of a wave moving through the sample during exposure by the electrons. This wave has a wavelength of about 0.02-0.03 Ångström (depending on the accelerating voltage of the electron microscope). Its phase is related to the phase of the undiffracted direct electron beam. The crystallographers, on the other hand, mean the "crystallographic structure factor phase" when they simply say "phase". This phase is the phase of standing waves of potential in the crystal (very similar to the electron density measured in X-ray crystallography). Each of these waves have their specific wavelength, called d-value for distance between so-called Bragg planes of low/high potential. These d-values range from the unit cell dimensions to the resolution limit of the electron microscope, i.e. typically from 10 or 20 Ångströms down to 1 or 2 Ångströms. Their phases are related to a fixed point in the crystal, defined in relation to the symmetry elements of that crystal. The crystallographic phases are a property of the crystal, so they exist also outside the electron microscope. The electron waves vanish if the microscope is switched off. In order to determine a crystal structure, it is necessary to know the crystallographic structure factors, but not to know the electron wave phases.

Just as with proteins, it has been possible to determine the atomic structures of inorganic crystals by electron crystallography. For simpler structure it is sufficient to use three perpendicular views, but for more complicated structures, also projections down ten or more different diagonals may be needed.

In addition to electron microscopy images, it is also possible to use electron diffraction (ED) patterns for crystal structure determination. The utmost care must be taken to record such ED patterns from the thinnest areas in order to keep most of the structure related intensity differences between the reflections (quasi-kinematical diffraction conditions). Just as with X-ray diffraction patterns, the important crystallographic structure factor phases are lost in electron diffraction patterns and must be uncovered by special crystallographic methods such as direct methods, maximum likelihood or (more recently) by the charge-flipping method. On the other hand, ED patterns of inorganic crystals have often a high resolution (= interplanar spacings with high Miller indices) much below 1 Ångström. This is comparable to the point resolution of the best electron microscopes. Under favourable conditions it is possible to use ED patterns from a single orientation to determine the complete crystal structure. Alternatively a hybrid approach can be used which uses HRTEM images for solving and intensities from ED for refining the crystal structure.

Recent progress for structure analysis by ED was made by introducing the Vincent-Midgley precession technique for recording electron diffraction patterns. The thereby obtained intensities are usually much closer to the kinematical intensities, so that even structures can be determined that are out of range when processing conventional (selected area) electron diffraction data.

Crystal structures determined via electron crystallography can be checked for their quality by using first-principles calculations within density functional theory (DFT). This approach was for the first time applied for the validation of several metal-rich structures which were only accessible by HRTEM and ED, respectively.

Recently, two very complicated zeolite structures have been determined by electron crystallography combined with X-ray powder diffraction. These are more complex than the most complex zeolite structures determined by X-ray crystallography.

Nuclear Magnetic Resonance Crystallography

Nuclear magnetic resonance crystallography (NMR crystallography) is a method which utilizes primarily NMR spectroscopy to determine the structure of solid materials on the atomic scale. Thus, solid-state NMR spectroscopy would be used primarily, possibly supplemented by quantum chemistry calculations (e.g. density functional theory), powder diffraction etc. If suitable crystals can be grown, any crystallographic method would generally be preferred to determine the crystal structure comprising in case of organic compounds the molecular structures and molecular packing. The main interest in NMR crystallography is in microcrystalline materials which are amenable to this method but not to X-ray, neutron and electron diffraction. This is largely because interactions of comparably short range are measured in NMR crystallography.

When applied to organic molecules, NMR crystallography aims at including structural information not only of a single molecule but also on the molecular packing (i.e. crystal structure). Contrary to X-ray, single crystals are not necessary with solid-state NMR and structural information can be obtained from high-resolution spectra of disordered solids. E.g. polymorphism is an area of interest for NMR crystallography since this is encountered occasionally (and may often be previously undiscovered) in organic compounds. In this case a change in the molecular structure and/or in the molecular packing can lead to polymorphism, and this can be investigated by NMR crystallography.

Dipolar Couplings-based Approach

The spin interaction that is usually employed for structural analyses via solid state NMR spectroscopy is the magnetic dipolar interaction. Additional knowledge about other interactions within the studied system like the chemical shift or the electric quadrupole interaction can be helpful as well, and in some cases solely the chemical shift has been employed as e.g. for zeolites. The "dipole coupling"-based approach parallels protein NMR spectroscopy to some extent in that e.g. multiple residual dipolar couplings are measured for proteins in solution, and these couplings are used as constraints in the protein structure calculation.

In NMR crystallography the observed spins in case of organic molecules would often be spin-1/2 nuclei of moderate frequency ($^{13}C, ^{15}N, ^{31}P$ etc.). i.e. 1H is excluded due to its large magnetogyric ratio and high spin concentration leading to a network of strong homonuclear dipolar couplings. There are two solutions with respect to $^{1}H :^{1} H$ spin diffusion experiments and specific labelling with ^{2}H spins (spin = 1). The latter is also popular e.g. in NMR spectroscopic investigations of hydrogen bonds in solution and the solid state. Both intra- and intermolecular structural elements can be investigated e.g. via deuterium REDOR (an established solid state NMR pulse sequence to measure dipolar couplings between deuterons and other spins). This can provide an additional constraint for an NMR crystallographic structural investigation in that it can be used to find and characterize e.g. intermolecular hydrogen bonds.

Dipolar Interaction

The above-mentioned dipolar interaction can be measured directly, e.g. between pairs of heteronuclear spins like $^{13}C/^{15}N$ in many organic compounds. Furthermore, the strength of the dipolar interaction modulates parameters like the longitudinal relaxation time or the spin diffusion rate which therefore can be examined to obtain structural information. E.g. ^{1}H spin diffusion has been measured providing rich structural information.

Chemical Shift Interaction

The chemical shift interaction can be used in conjunction with the dipolar interaction to determine the orientation of the dipolar interaction frame (principal axes system) with respect to the molecular frame (dipolar chemical shift spectroscopy). For some cases there are rules for the chemical shift interaction tensor orientation as for the ^{13}C spin in ketones due to symmetry arguments (sp^2 hybridisation). If the orientation of a dipolar interaction (between the spin of interest and e.g. another heteronucleus) is measured with respect to the chemical shift interaction coordinate system, these two pieces of information (chemical shift tensor/molecular orientation and the dipole tensor/chemical shift tensor orientation) combined give the orientation of the dipole tensor in the molecular frame. However, this method is only suitable for small molecules (or polymers with a small repetition unit like polyglycine) and it provides only selective (and usually intramolecular) structural information.

Crystal Structure Refinements

The dipolar interaction yields the most direct information with respect to structure as it makes it possible to measure the distances between the spins. The sensitivity of this interaction is however lacking and even though dipolar-based NMR crystallography makes the elucidation of structures possible, other methods are necessary to obtain high resolution structures. For these reasons much work was done to include the use other NMR observables such as chemical shift anisotropy, J-coupling and the quadrupolar interaction. These anisotropic interactions are highly sensitive to the 3D local environment making it possible to refine the structures of powdered samples to structures rivaling the quality of single crystal X-ray diffraction. These however rely on adequate methods for predicting these interactions as they do not depend in a straightforward fashion on the structure.

Comparison with Diffraction Methods

A drawback of NMR crystallography is that the method is typically more time consuming and more expensive (due to spectrometer costs and isotope labelling) than X-ray crystallography, it often elucidates only part of the structure, and isotope labelling and experiments may have to be tailored to obtain key structural information. Also not always is a molecular structure suitable for a pure NMR-based NMR crystallographic approach, but it can still play an important role in a multimodality (NMR+diffraction) study.

Unlike in the case of diffraction methods, it appears that NMR crystallography needs to work on a case by case basis. This is the case since difference systems will have different spin physics and different observables which can be probed. The method may therefore not find widespread use as different systems will require qualified individuals to design experiments to study them.

Time Resolved Crystallography

Time-resolved crystallography is an experimental technique that can detect molecular changes at atomic and temporal resolutions. Due to the feasibility of rapid reaction initiation by a laser pulse, this technique has been widely used to study light-active protein systems including myoglobin, hemoglobin,photoactive yellow protein, photosystem II, and rhodopsin. Recent advances in sample delivery systems and femtosecond X-ray free electron lasers (XFELs) have allowed time-resolved serial femtosecond crystallography (TR-SFX) to be extended to other systems as well. For example, the mix-and-inject method can rapidly and uniformly initiate an enzymatic reaction or an RNA-ligand interactionin micro/nanocrystals before diffraction whose rate is limited by diffusion. To the best of our knowledge, time-resolved crystallography has almost exclusively been applied on time scales of picoseconds to seconds. Here, using traditional cryocrystallography at a synchrotron source without rapid mixing or use of a laser pulse, we recently discovered the slow oxidative decay of tetrahydrofolate to dihydrofolate in the enzyme bound crystalline form at room temperature, with the critical transition occurring 2–3 days after the crystallization setup. We present an analysis of the third day's crystal structure to estimate the putative intermediate's geometry and population relative to dihydrofolate and tetrahydrofolate.

Materials and Methods

Protein Expression and Purification

C-terminal 6xHis-tagged *Escherichia coli* DHFR (dihydrofolate reductase, with 100% sequence identity to UNIPROT sequence sp|P0ABQ4| or sp|P0ABQ5|) was generously provided by Drs. Eugene Shakhnovich and João Rodrigues. DHFR was overexpressed in *E. coli* BL21 and then purified by Ni-NTA and size exclusion chromatography. The initial protein stock was stored at −80 °C at a concentration of 30 mg ml^{-1} in 20 mM Tris, at pH 8, and in 1 mM dithiothreitol (DTT).

Protein Crystallization

The current intermediate time point complex was obtained using the same crystal growth condition as in our previously reported binary tetrahydrofolate complex (PDB: 6CW7). Briefly, as-purified eDHFR was crystallized by sitting drop vapor diffusion using a 1:1 v/v mixing of 20 mg ml^{-1} DHFR solution in 13.3 mM Tris at pH 8, 16.7 mM HEPES at pH 7.3, 33.3 mM NaCl, and 0.67 mM DTT with the reservoir solution containing 0.1 M MES, at pH 6.5, with 30% w/v PEG 3350, and 0.4 M $MgCl_2$. Mixed drops of 0.8 μl were equilibrated over a reservoir solution of 50 μl on a MRC 2-well plate and incubated at 20 °C in the dark. A single crystal per dataset was harvested at 3 days after setting up the crystallization drops and then cryoprotected with LV CryoOil and flash-frozen in liquid N_2.

Data Collection and Refinement

Diffraction data were collected at the Advanced Photon Source at Argonne National Laboratory on the LRL-CAT (31-ID-D) beamline at 100 K. The detector was a Rayonix 225 HE CCD (Rayonix) using a single wavelength of 0.97931 Å. The intermediate time point (3 days of crystal growth) dataset was collected and processed to a resolution of 1.35 Å. The datasets were indexed, integrated, and scaled using XDS. The structure was determined by molecular replacement with Phaser_MR (using the protein coordinates of the tetrahydrofolate complex as the search model) and completed by alternating rounds of manual model building with COOT and both reciprocal and real space refinements using phenix.refine of the PHENIX suite. The ligand was originally built into the model as tetrahydrofolate based on the apparent tetrahedral geometry suggested by the clear omit electron density. Met16 is very close in geometry to that in the tetrahydrofolate complex. We conjecture that it acts as a clamp that is partly responsible for the slowdown of tetrahydrofolate to dihydrofolate oxidation. The Met20 loop appears to be partially disordered for residues Glu17-Asn18-Ala19 in contrast to the clearly traceable conformations of the end-state complexes with tetrahydrofolate or dihydrofolate. On the other hand, Met20 and Pro21 are clearly visible and are quite close to their conformation in the dihydrofolate complex. Thus, the residues in the Met20 loop adopt structure of the intermediate in complex with DHFR are a hybrid of the two endpoint conformations with disorder in the middle residues of the loop. Mg^{2+}, Cl^{-1} ions, and water molecules were added to the model after the ligand was built in.

The fully refined model contains a ligand geometry containing coplanar pterin rings and a tetrahedral C6 geometry suggesting a possible intermediate state. This contrasts with the puckered pterin previously observed for the tetrahydrofolate complex despite the same ligand restraints being used. Due to the uncertainty of the exact chemical structure and fractional population of the ligand, we further performed refinement analysis by comparing one-state (a single intermediate), two-state (a linear combination of two end states, dihydrofolate and tetrahydrofolate), three-state (a linear combination of both end states and the proposed intermediate) models. Briefly, the previously reported end-state models (PDB IDs: 6CW7 and 6CXK) were each aligned to the current protein model and their ligand coordinates extracted and merged into the current model with COOT. To achieve the linear combination of ligand states, only the occupancy and individual B-factors of the structures were refined in reciprocal space with their coordinates kept intact. The initial ligand occupancy was assigned as 0.5 to each ligand for the two-state model. In the three-state model, the initial ligand occupancy was intentionally assigned as 0.1 to the proposed intermediate state and 0.45 to each of the two end-state ligands to test how these three states compete with each other in terms of percentage population in the current refinement scheme. The final refined occupancy values of the three ligands appear to be insensitive to their initial occupancy assignments (e.g., equal initial occupancy of 0.33 for each ligand state) and always converged to the same results, i.e., the intermediate state is the most populated at 0.45 which approximately equals the combined occupancy of two-end states. The structures determined in this study display Ramachandran statistics absent of outliers, with 98.8% of the residues in the most favored regions and 1.2% of residues in additionally allowed regions of the Ramachandran diagram defined by MolProbity. All structures are displayed using PyMOL unless otherwise stated.

Bayesian Difference Refinement

Bayesian difference refinement was performed using the default protocol developed by Terwilliger and Berendzen. This is a sensitive method to detect small but finite structural changes between two very similar protein structural models obtained from two very similar experimental X-ray diffraction datasets. It was reported to perform better or at least as good as individual refinements in estimating the finite atomic shifts (RMS of shifts), depending on the correlation coefficient of model errors between test datasets. The more similar the two datasets (hence the models) are, the more useful is the Bayesian difference refinement method. Briefly, a pseudo variant dataset with amplitudes and weights was generated by running FDIFF scripts in the SOLVE program based on *F*o (the observed amplitude of the structure factor) and *F*c (the amplitude calculated from the model) from both native and variant datasets. The native dataset was the tetrahydrofolate complex (PDB: 6CW7), and the variant dataset was the intermediate time point dataset. Both are isomorphous crystals with nearly identical unit cell parameters grown under the same crystallization conditions, with around a 24 h delay in harvesting (2 days vs. 3 days of crystal growth since setup).

The parameters of Bayesian difference refinement were defined originally by Terwilliger and Berendzen as follows:

The value of F_{diff} is given by,

$$F_{diff} = F_{c,native} + \beta * \left(F_{o,variant} - F_{o,native}\right).$$

The factor β, essentially the correlation coefficient between $Fo_{native}-Fc_{native}$ and $Fo_{variant}-Fc_{variant}$, is given by,

$$\beta = E^2 / \left(E^2 + A_{native}{}^2 + \sigma_{native}{}^2\right).$$

The weighting factor or pseudo experimental errors of the pseudo dataset is given by,

$$\sigma_{Fdiff}{}^2 = \sigma_{variant}{}^2 + A_{variant}{}^2 + 1/\left[1/\left(\sigma_{native}{}^2\ A_{native}{}^2\right)\right.$$

E^2 represents the total correlated model error vs. the data between the native and variant datasets. $A_{native}{}^2$ and $A_{variant}{}^2$ represent uncorrelated model errors of the native and variant datasets, respectively. $\sigma_{native}{}^2$ and $\sigma_{variant}{}^2$ represent the experimental measurement errors of the native and variant datasets, respectively. Both β and $\sigma_{Fdiff}{}^2$ terms contain certain weighted information derived from errors in both models and the experimental measurement of the two original datasets. The pseudo dataset containing information of F_{diff} (amplitudes) and $\sigma_{Fdiff}{}^2$ (weights) can then be used as input for the structure factor data in phenix.refine following regular refinement procedures.

References

- Scapin G (2006). "Structural biology and drug discovery". Current Pharmaceutical Design. 12 (17): 2087–97. doi:10.2174/138161206777585201. PMID 16796557
- Robinson, Philip (26 February 2009). "Crystal clear method for identifying powders". Highlights in Chemical Technology. Retrieved 2015-10-22
- Markus-Neuburger-PhD-thesis-edoc: edoc.unibas.ch, Retrieved 17 July, 2020
- Beatrice Roessler, Sven Kleinhenza, Konrad Seppelt (2000). "Pentamethylmolybdenum". Chemical Communications (12): 1039–1040. doi:10.1039/b000987n
- "WIDE-ANGLE X-RAY DIFFRACTION THEORY VERSUS CLASSICAL DYNAMICAL THEORY" by S.G. Podorov, A. Nazarkin, Recent Res. Devel. Optics, 7 (2009) ISBN 978-81-308-0370-8

3

Crystal Structure

Crystal structure is the ordered arrangement of atoms, molecules and ions in a crystal. It is used in materials science to characterize different materials and phase identification. All the aspects related to crystal structure have been carefully analyzed in this chapter.

In mineralogy and crystallography, a crystal structure is a unique arrangement of atoms in a crystal.

Crystal Lattice

Crystal Lattice is a three-dimensional representation of atoms and molecules arranged in a specific order/pattern. In other words, a crystal lattice can be defined as a geometrical arrangement of constituent particles of matter (atoms, ions or molecules) as points in space. There are total 14 possible three-dimensional lattices. Crystal lattices are also known by Bravais Lattices, named after the scientist Auguste Bravais.

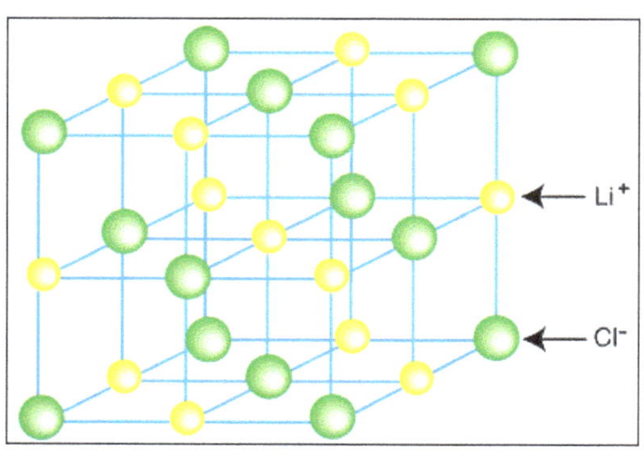

Example of a crystal lattice.

Characteristics of Crystal Lattices

The following characteristics are depicted by Bravais lattices:

- Each point in a lattice represents a lattice site or we can say lattice point.

- Each point denotes a particular type of constituent particles of matter be it an atom, molecule or an ion.
- By joining the lattices points inside the lattice we can define geometry of the lattice.

Unit Cell

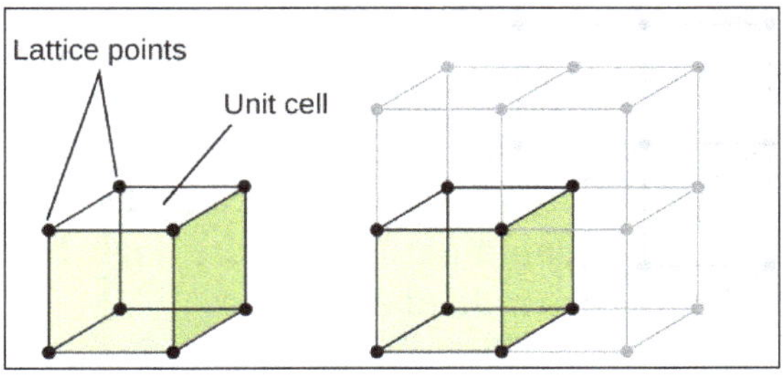

A unit cell is the shortest portion of a lattice.

The smallest possible portion or part of the crystal lattice which repeats itself in different directions of the lattice is called the unit cell. Many unit cells combine to geometrically form the crystal lattice.

Characteristics of Unit Cell

The following characteristics define a unit cell:

- A unit cell has three edges a, b and c and three angles α, β and γ between the respective edges.
- The a, b and c may or may not be mutually perpendicular.
- The angle between edge b and c is α, a and c is β and that of between a and b is γ.

Formation of Lattice

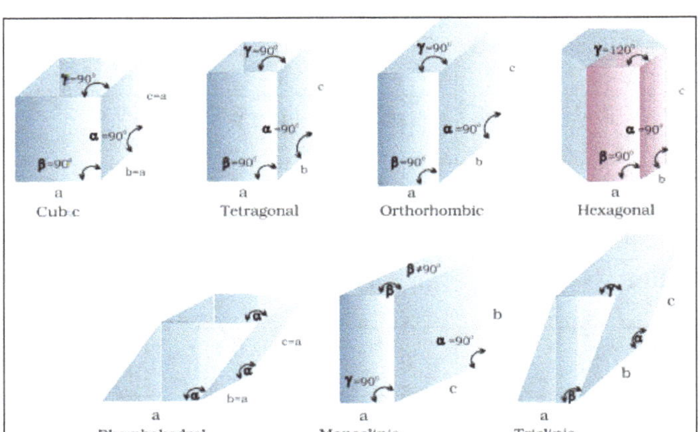

Types of unit cells which form crystal lattices.

Formation of unit cells takes place in seven forms, namely:

- Cubic Lattice: Cubic lattice is formed into three possible geometries of unit cells: primitive, body-centred and face centred unit cells. In a cubic lattice, all the edges are equal and the angle between their faces is 90° that is, mutually perpendicular.
- Tetragonal Lattice: The formation of tetragonal lattice takes place in two geometries of unit cells: primitive and body centred unit cells. In a tetragonal lattice, only one edge has different length and angle between respective edges is 90° that is, mutually perpendicular.
- Orthorhombic Lattice: There are four types of orthorhombic lattice mainly: primitive, end-centred, body-centred and face centred. In orthorhombic lattice, the edge lengths are unequal in nature and the angle between respective edges is 90° that is, mutually perpendicular.
- Monoclinic Lattice: Monoclinic lattice is formed from two types of unit cells namely: primitive and end centred. Monoclinic lattice has unequal sides and two angles between the faces of the lattice are 90°.
- Hexagonal Lattice: Hexagonal lattice is formed from only one type of unit cell that is, primitive. In hexagonal lattice, only one side and two angles are 90° and one angle is 120°.
- Rhombohedral Lattice: Rhombohedral Lattice is also formed from one type of unit cell that is, primitive. In Rhombohedral Lattice all the sides are equal and two angles between the faces of the rhombohedral lattice are less than 90°.
- Triclinic Lattice: The formation of triclinic lattice also takes place from one type of unit cell that is, primitive. In triclinic lattice all the sides are unequal and none of the angles between the faces of the triclinic lattice are 90°.

Table: Types of lattice formation.

Lattice	Types	Edge Length	Angles between faces	Examples
Cubic	Primitive, Body-centred, Face-centred	$a = b = c$	$\alpha = \beta = \gamma = 90^\circ$	NaCl, Copper and ZnS
Tetragonal	Primitive, Body-centred	$a = b \neq c$	$\alpha = \beta = \gamma = 90^\circ$	White tin, SnO_2, TiO_2 and $CaSO_4$
Orthorhombic	Primitive, Body-centred, Face-centred , End-centred	$a \neq b \neq c$	$\alpha = \beta = \gamma = 90^\circ$	Rhombic Sulphur, $BaSO_4$and KNO_3
Hexagonal	Primitive	$a = b \neq c$	$\alpha = \beta = 90^\circ$ and $\gamma = 120^\circ$	Graphite, ZnO and CdS
Rhombohedral	Primitive	$a = b = c$	$\alpha = \beta = \gamma \neq 90^\circ$	$CaCO_3$ (Calcite) and HgS (cinnabar)
Monoclinic	Primitive, End-centred	$a \neq b \neq c$	$\alpha = \gamma = 90^\circ$ $\beta \neq 90^\circ$	Sulphur
Triclinic	Primitive	$a \neq b \neq c$	$\alpha \neq \beta \neq \gamma \neq 90^\circ$	H_3PO_3 , $CuSO_4.5\ H_2O$

Crystal Habit

Pyrite sun (or dollar) in laminated shale matrix. Between tightly spaced layers of shale, the aggregate was forced to grow in a laterally compressed, radiating manner. Under normal conditions, pyrite would form cubes or pyritohedrons.

In mineralogy, crystal habit is the characteristic external shape of an individual crystal or crystal group. A single crystal's habit is a description of its general shape and its crystallographic forms, plus how well developed each form is.

Recognizing the habit may help in identifying a mineral. When the faces are well-developed due to uncrowded growth a crystal is called *euhedral*, one with partially developed faces is *subhedral*, and one with undeveloped crystal faces is called *anhedral*. The long axis of a euhedral quartz crystal typically has a six-sided prismatic habit with parallel opposite faces. Aggregates can be formed of individual crystals with euhedral to anhedral grains. The arrangement of crystals within the aggregate can be characteristic of certain minerals. For example, minerals used for asbestos insulation often grow in a fibrous habit, a mass of very fine fibers.

The terms used by mineralogists to report crystal habits describe the typical appearance of an ideal mineral. Recognizing the habit can aid in identification as some habits are characteristic. Most minerals, however, do not display ideal habits due to conditions during crystallization. Euhedral crystals formed in uncrowded conditions with no adjacent crystal grains are not common; more often faces are poorly formed or unformed against adjacent grains and the mineral's habit may not be easily recognized.

Goethite replacing pyrite cubes.

Factors influencing habit include: a combination of two or more crystal forms; trace impurities present during growth; crystal twinning and growth conditions (i.e., heat, pressure, space); and specific growth tendencies such as growth striations. Minerals belonging to the same crystal system do not necessarily exhibit the same habit. Some habits of a mineral are unique to its variety and locality. For example, while most sapphires form elongate barrel-shaped crystals, those found in Montana form stout *tabular* crystals. Ordinarily, the latter habit is seen only in ruby. Sapphire and ruby are both varieties of the same mineral: corundum.

Some minerals may replace other existing minerals while preserving the original's habit: this process is called pseudomorphous replacement. A classic example is tiger's eye quartz, crocidolite asbestos replaced by silica. While quartz typically forms *prismatic* (elongate, prism-like) crystals, in tiger's eye the original *fibrous* habit of crocidolite is preserved.

The names of crystal habits are derived from:

- Predominant crystal faces (prism – prismatic, pyramid – pyramidal and pinacoid – platy).
- Crystal forms (cubic, octahedral, dodecahedral).
- Aggregation of crystals or aggregates (fibrous, botryoidal, radiating, massive).
- Crystal appearance (foliated/lamellar (layered), dendritic, bladed, acicular, lenticular, tabular (tablet shaped)).

Direct Lattice

The direct lattice represents the triple periodicity of the ideal infinite perfect periodic structure that can be associated to the structure of a finite real crystal. To express this periodicity one calls crystal pattern an object in point space E^n (direct space) that is invariant with respect to three linearly independent translations, $\mathbf{t}_1$, $\mathbf{t}_2$ and $\mathbf{t}_3$. One distinguishes two kinds of lattices, the vector lattices and the point lattices.

Any translation $\mathbf{t} = u^i\,\mathbf{t}_i$ (u^i arbitrary integers) is also a translation of the pattern and the infinite set of all translation vectors of a crystal pattern is the vector lattice **L** of this crystal pattern.

Given an arbitrary point P in point space, the set of all the points P_i deduced from one of them by a translation $\mathbf{PP}_i = \mathbf{t}_i$ of the vector lattice **L** is called the point lattice.

A basis **a**, **b**, **c** of the vector space $\mathbf{V}^n$ is a crystallographic basis of the vector lattice **L** if every integral linear combination $\mathbf{t} = u\,\mathbf{a} + v\,\mathbf{b} + w\,\mathbf{c}$ is a lattice vector of **L**. It is called a primitive basis if every lattice vector **t** of **L** may be obtained as an integral linear combination of the basis vectors, **a**, **b**, **c**. Referred to any crystallographic basis the coefficients of each lattice vector are either integral or rational, while in the case of a primitive basis they are integral. Non-primitive bases are used conventionally to describe centred lattices.

The parallelepiped built on the basis vectors is the unit cell. Its volume is given by the triple scalar product, V = (**a**, **b**, **c**).

If the basis is primitive, the unit cell is called the primitive cell. It contains only one lattice point. If the basis is non-primitive, the unit cell is a multiple cell and it contains more than one lattice point. The multiplicity of the cell is given by the ratio of its volume to the volume of a primitive cell.

Direct Space

The direct space (or *crystal space*) is the *point space*, E^n, in which the structures of finite real crystals are idealized as infinite perfect three-dimensional structures. To this space one associates the *vector space*, V^n, of which lattice and translation vectors are elements. It is a *Euclidean* space where the scalar product of two vectors is defined. The two spaces are connected through the following relations:

- To any two points P and Q of the point space E^n a vector **PQ** = **r** of the vector space Vn is attached.
- For each point P of E^n and for each vector r of V^n there is exactly one point Q of En for which **PQ** = **r** holds.
- If R is a third point of the point space, **PQ** + **QR** = **PR**.

Reciprocal Lattice

The reciprocal lattice is constituted of the set of all possible linear combinations of the basis vectors **a***, **b***, **c*** of the reciprocal space. A point (*node*), *H*, of the reciprocal lattice is defined by its position vector:

$$\mathbf{OH} = \mathbf{r}^*_{\mathbf{hkl}} = h\mathbf{a}^* + k\mathbf{b}^* + /\mathbf{c}^*.$$

If *H* is the *n*th node on the row *OH*, one has:

$$\mathbf{OH} = n\,\mathbf{OH}_1 = n\,(h_1\mathbf{a}^* + k_1\mathbf{b}^* + /_1\mathbf{c}^*),$$

where H_1 is the first node on the row *OH* and h_1, k_1, l_1 are relatively prime.

Geometrical Applications

Each vector $\mathbf{OH} = \mathbf{r}^*_{\mathbf{hkl}} = h\,\mathbf{a}^* + k\,\mathbf{b}^* + l\,\mathbf{c}^*$ of the reciprocal lattice is associated with a family of direct lattice planes. It is normal to the planes of the family, and the lattice spacing of the family is $d = 1/OH_1 = n/OH$ if H is the nth node on the reciprocal lattice row OH. One usually sets $d_{hkl} = d/n = 1/OH$. If $\mathbf{OP} = x\,\mathbf{a} + y\,\mathbf{b} + z\,\mathbf{c}$ is the position vector of a point of a lattice plane, the equation of the plane is given by $\mathbf{OH_1} \cdot \mathbf{OP} = K$ where K is a constant integer. Using the properties of the scalar product of a reciprocal space vector and a direct space vector, this equation is $\mathbf{OH_1} \cdot \mathbf{OP} = h_1 x + k_1 y + l_1 z = K$. The Miller indices of the family are h_1, k_1, l_1. The subscripts of the Miller indices will be dropped hereafter.

The Miller indices of the family of lattice planes parallel to two direct space vectors, $\mathbf{r_1} = u_1\,\mathbf{a} + v_1\,\mathbf{b} + w_1\,\mathbf{c}$ and $\mathbf{r_2} = u_2\,\mathbf{a} + v_2\,\mathbf{b} + w_2\,\mathbf{c}$ are proportional to the coordinates in reciprocal space, h, k, l, of the vector product of these two vectors:

$$h/(v_1 w_2 - v_2 w_1) = k/(w_1 u_2 - w_2 u_1) = l/(u_1 v_2 - u_2 v_1).$$

The coordinates u, v, w in direct space of the zone axis intersection of two families of lattice planes of Miller indices h_1, k_1, l_1 and h_2, k_2, l_2, respectively, are proportional to the coordinates of the vector product of the reciprocal lattice vectors associated with these two families:

$$u/(k_1 l_2 - k_2 l_1) = v/(l_1 h_2 - l_2 h_1) = w/(h_1 k_2 - h_2 k_1).$$

Centred Lattices

Direct lattice			Reciprocal lattice		
Bravais letter	Centring vectors	Unit-cell volume V_c	Bravais letter	Multiple unit cell	Unit cell volume V^*_c
P	0	V	P	$\mathbf{a}^*_c, \mathbf{b}^*_c, \mathbf{c}^*_c$	V^*
A	$½\mathbf{b}_c + ½\mathbf{c}_c$	$2V$	A	$\mathbf{a}^*_c, 2\mathbf{b}^*_c, 2\mathbf{c}^*_c$	$½V^*$
B	$½\mathbf{c}_c + ½\mathbf{a}_c$	$2V$	B	$2\mathbf{a}^*_c, \mathbf{b}^*_c, 2\mathbf{c}^*_c$	$½V^*$
C	$½\mathbf{a}_c + ½\mathbf{b}_c$	$2V$	C	$2\mathbf{a}^*_c, 2\mathbf{b}^*_c, \mathbf{c}^*_c$	$½V^*$
I	$½\mathbf{a}_c + ½\mathbf{b}_c + ½\mathbf{c}_c$	$2V$	F	$2\mathbf{a}^*_c, 2\mathbf{b}^*_c, 2\mathbf{c}^*_c$	$½V^*$
F	$½\mathbf{a}_c + ½\mathbf{b}_c$ $½\mathbf{b}_c + ½\mathbf{c}_c$ $½\mathbf{c}_c + ½\mathbf{a}_c$	$4V$	I	$2\mathbf{a}^*_c, 2\mathbf{b}^*_c, 2\mathbf{c}^*_c$	$¼V^*$
R (rhombohedral axes)	0	V	R	$\mathbf{a}^*_c, \mathbf{b}^*_c, \mathbf{c}^*_c$	V^*
R (hexagonal axes)	$⅔\mathbf{a}^*_c + ⅓\mathbf{b}^*_c + ⅓\mathbf{c}^*_c$ $⅓\mathbf{a}^*_c + ⅔\mathbf{b}^*_c + ⅔\mathbf{c}^*_c$	$3V$	R	$3\mathbf{a}^*_c, 3\mathbf{b}^*_c, 3\mathbf{c}^*_c$	$⅓V^*$

where $\mathbf{a}_c, \mathbf{b}_c, \mathbf{c}_c$ are the basis vectors of the conventional multiple cell and $\mathbf{a}^*_c, \mathbf{b}^*_c, \mathbf{c}^*_c$ the corresponding reciprocal lattice vectors.

Diffraction Condition in Reciprocal Space

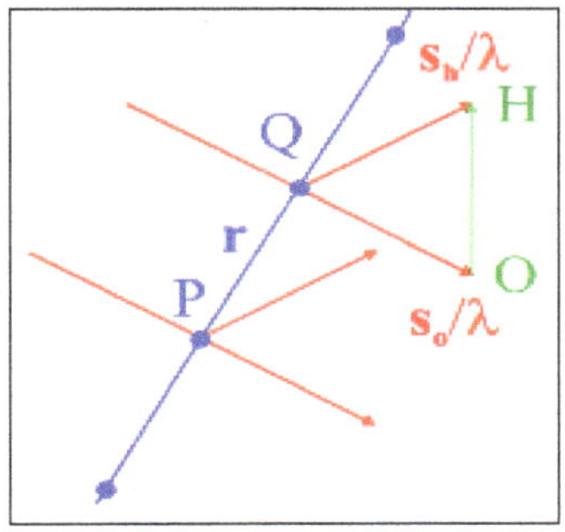

The condition that the waves outgoing from two point scatterers separated by a lattice vector $\mathbf{r} = u\,\mathbf{a} + v\,\mathbf{b} + w\,\mathbf{c}$ (u, v, w integers) be in phase is that the scalar product $(\mathbf{s}_h/\lambda - \mathbf{s}_o/\lambda) \cdot \mathbf{r}$, where $\mathbf{s}_h$ and $\mathbf{s}_o$ are unit vectors in the scattered and incident directions, respectively, be an integer, n. This condition is satisfied whatever $\mathbf{r}$ if the diffraction vector ($\mathbf{OH} = \mathbf{s}_h/\lambda - \mathbf{s}_o/\lambda$) is of the form:

$$(\mathbf{s}_h / \lambda - \mathbf{s}_o / \lambda) = h\mathbf{a}^* + k\mathbf{b}^* + l\mathbf{c}^*,$$

where h, k, l are integers, namely the diffraction vector **OH** is a vector of the reciprocal lattice.

A node of the reciprocal lattice is therefore associated with each Bragg reflection on the lattice planes of Miller indices $(h/K, k/K, l/K)$. It is called the *hkl* reflection.

The relation $\mathbf{s}_h/\lambda - \mathbf{s}_o/\lambda = \mathbf{OH}$ generalizes the Laue equations. It is equivalent to Bragg's law.

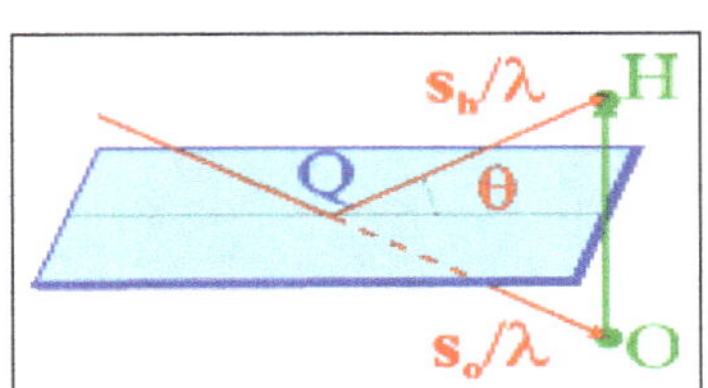

Consider the lattice plane passing through lattice point Q and perpendicular to reciprocal-lattice vector **OH** and let θ be the angle between the incident, $\mathbf{s}_o$, or the reflected, $\mathbf{s}_h$, directions and the lattice plane. It can be seen from the figure that:

$$OH / 2 = \sin \theta / \lambda,$$

and, since $OH = n/d$ (d lattice spacing of the family of lattice planes associated with **OH**) and $d_{hkl} = d/n$:

$$2d \sin \theta = n\lambda, \text{ or } 2d_{hkl} \sin \theta = \lambda,$$

which is Bragg's law. n is the order of the reflection.

Another way to express the diffraction condition in reciprocal space is to consider a sphere centred at a node Q of the direct lattice, of radius $1/\lambda$ and passing through the origin O of the reciprocal lattice. If it passes through another node, H, of the reciprocal lattice, Bragg's law is satisfied for the family of direct lattice planes associated with that node and of lattice spacing $d_{hkl} = n/OH$ if H is the nth node on the row **OH**. This sphere is called the Ewald sphere.

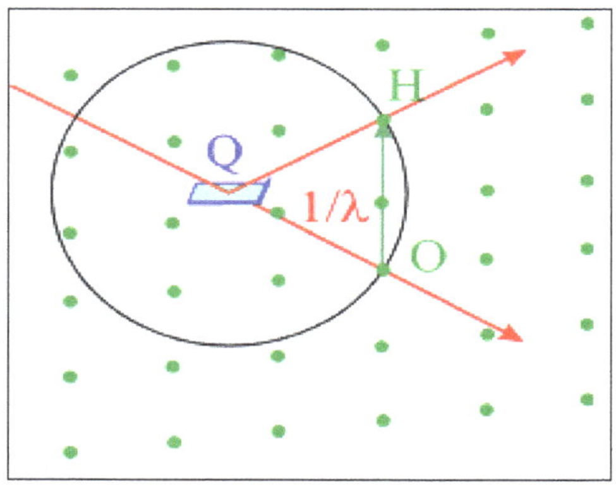

$Cr_{23}C_6$ Crystal Structure

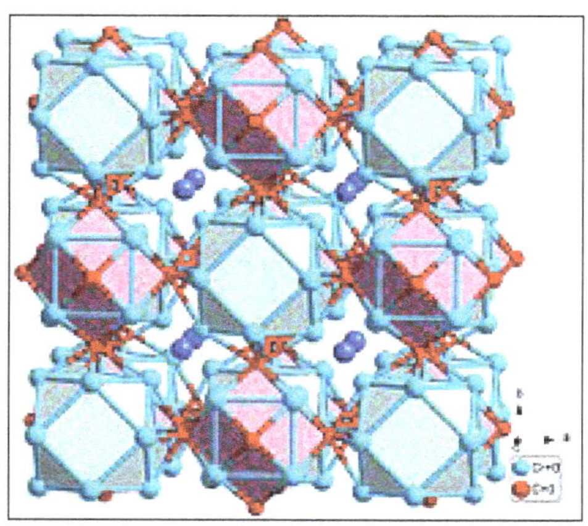

The $Cr_{23}C_6$ structure represented as polyhedra of atoms. The blue surfaces outline cuboctahedra of chromium atoms, whereas the red surfaces outline chromium cubes capped by carbon atoms. The darker blue spheres represent chromium atoms outside of the polyhedra.

$Cr_{23}C_6$ is the prototypical compound of a common crystal structure, discovered in 1933 as part of the chromium-carbon binary phase diagram. Over 85 known compounds adopt this structure type, which can be described as a NaCl-like packing of chromium cubes and cuboctahedra.

Structure

The space group of this structure is called F m $\bar{3}$ m (in Hermann-Mauguin notation) or "225" (in the International Tables for Crystallography). It belongs to a cubic crystal system, with Pearson symbol cF116.

The shortest interatomic distances are between carbon and chromium atoms, which is expected on the basis of atomic size. The carbon atoms are in positions that cap each face of the chromium cubes and their coordination environment can be thought of as distorted square antiprisms formed from chromium atoms of both the cubes and the cuboctahedra.

The closest Cr-Cr contacts are between members of a cuboctahedron, and the third closest are between members of a cube. The members of the cube, however, are closer to the 8 chromium atoms in the unit cell that are not part of either polyhedron. The coordination environment of these other atoms can be thought of as distorted Friauf polyhedra composed of chromium atoms, if next-nearest neighbors are included.

Materials

Examples of compounds that form in this structure type include $Cr_{23}C_6$, $Mn_{23}C_6$, and many ternary intermetallic carbides and borides. A few phases of ternary silicides, germanides, and phosphides are also known to exist. In going from the binary to ternary systems, some of the transition metal atoms are substituted by a third element, which can be an alkali metal, alkaline earth metal, rare-earth element, main group element, or another transition metal. This leads to an empirical formula of the form $A_{23-x}B_xC_6$. Materials of this kind continue to be studied for potentially interesting magnetic and physical properties.

Crystal Structure Prediction

Crystal structure prediction (CSP) is the calculation of the crystal structures of solids from first principles. Reliable methods of predicting the crystal structure of a compound, based only on its composition, has been a goal of the physical sciences since the 1950s. Computational methods employed include simulated annealing, evolutionary algorithms, distributed multipole analysis, random sampling, basin-hopping, data mining, density functional theory and molecular mechanics.

The crystal structures of simple ionic solids have long been rationalised in terms of Pauling's rules, first set out in 1929 by Linus Pauling. For metals and semiconductors one has different rules involving valence electron concentration. However, prediction and rationalization are rather different things. Most commonly, the term crystal structure prediction means a search for the minimum-energy arrangement of its constituent atoms (or, for molecular crystals, of its molecules) in space. The problem has two facets: combinatorics (the "search phase space", in practice most acute for inorganic crystals), and energetics (or "stability ranking", most acute for molecular organic crystals). For complex non-molecular crystals (where the "search problem" is most acute), major recent advances have been the development of the Martonak version of metadynamics,, the Oganov-Glass evolutionary algorithm USPEX, and first principles random search. The latter are capable of solving the global optimization problem with up to around a hundred degrees of freedom, while the approach of metadynamics is to reduce all structural variables to a handful of "slow" collective variables (which often works).

Molecular Crystals

Predicting organic crystal structures is important in academic and industrial science, particularly for pharmaceuticals and pigments, where understanding polymorphism is beneficial. The crystal structures of molecular substances, particularly organic compounds, are very hard to predict and rank in order of stability. Intermolecular interactions are relatively weak and non-directional and long range. This results in typical lattice and free energy differences between polymorphs that are often only a few kJ/mol, very rarely exceeding 10 kJ/mol. Crystal structure prediction methods often locate many possible structures within this small energy range. These small energy differences are challenging to predict reliably without excessive computational effort.

Since 2007, significant progress has been made in the CSP of small organic molecules, with several different methods proving effective. The most widely discussed method first ranks the energies of all possible crystal structures using a customised MM force field, and finishes by using a dispersion-corrected DFT step to estimate the lattice energy and stability of each short-listed candidate structure. More recent efforts to predict crystal structures have focused on estimating crystal free energy by including the effects of temperature and entropy in organic crystals using vibrational analysis or molecular dynamics.

Crystal Structure Prediction Software

The following codes can predict stable and metastable structures given chemical composition and external conditions (pressure, temperature):

- AIRSS: Ab Initio Random Structure Searching based on stochastic sampling of configuration space and with the possibility to use symmetry, chemical, and physical constraints. Has been used to study bulk crystals, low-dimensional materials, clusters, point defects, and interfaces. Released under the GPL2 licence. Regularly updated.
- CALYPSO: The Crystal structure AnaLYsis by Particle Swarm Optimization, implementing the particle swarm optimization (PSO) algorithm to identify/determine the crystal structure. As with other codes, knowledge of the structure

can be used to design multi-functional materials (e.g.,superconductive, thermoelectric, superhard, and energetic materials). Free for academic researchers. Regularly updated.

- GASP: predicts the structure and composition of stable and metastable phases of crystals, molecules, atomic clusters and defects from first-principles. Can be interfaced to other energy codes including: VASP, LAMMPS, MOPAC, Gulp, JDFTx etc. Free to use and regularly updated.
- GRACE: for predicting molecular crystal structures, especially for the pharmaceutical industry. Based on dispersion-corrected density functional theory. Commercial software under active development.
- GULP: Monte Carlo and genetic algorithms for atomic crystals. GULP is based on classical force fields and works with many types of force fields. Free for academic researchers. Regularly updated.
- USPEX: Multi-method software that includes evolutionary algorithms and other methods (random sampling, evolutionary metadynamics, improved PSO, variable-cell NEB method and transition path sampling method for phase transition mechanisms). Can be used for atomic and molecular crystals; bulk crystals, nanoparticles, polymers, surface reconstructions, interfaces; can optimize the energy or other physical properties. In addition to finding the structure for a given composition, can identify all stable compositions in a multicomponent variable-composition system and perform simultaneous optimisation of several properties. Free for academic researchers. Used by >4500 researchers. Regularly updated.
- XtalOpt: Open source code implementing an evolutionary algorithm.

Crystal Structure of Boron-rich Metal Borides

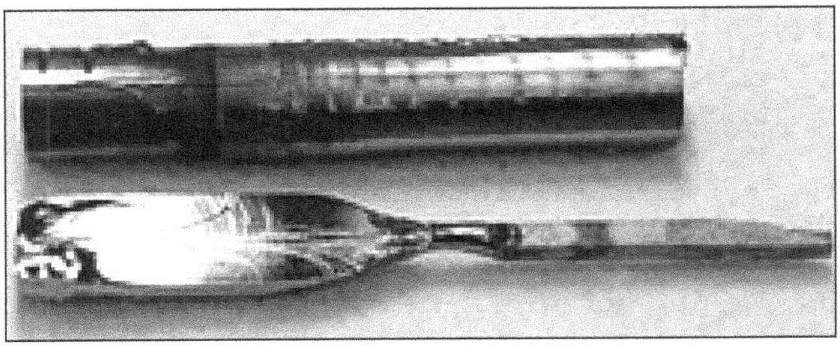

Two single crystals of YB_{66} (1 cm diameter) grown by floating zone technique using (100) oriented seeds. In the top crystal, the seed (left from the black line) has same diameter as the crystal. In the bottom crystal (sliced), the seed is much thinner and is on the right side.

Metals, and specifically rare-earth elements, form numerous chemical complexes with boron. Their crystal structure and chemical bonding depend strongly on the metal element M and on its atomic ratio to boron. When B/M ratio exceeds 12, boron atoms form B_{12} icosahedra which are linked into a three-dimensional boron framework, and the metal atoms reside in the voids of this framework. Those icosahedra are basic structural units of most allotropes of boron and boron-rich rare-earth borides. In such borides, metal atoms donate electrons to the boron polyhedra, and thus these compounds are regarded as electron-deficient solids.

The crystal structures of many boron-rich borides can be attributed to certain types including $MgAlB_{14}$, YB_{66}, $REB_{41}Si_{1.2}$, B_4C and other, more complex types such as $RE_xB_{12}C_{0.33}Si_{3.0}$. Some of these formulas, for example B_4C, YB_{66} and $MgAlB_{14}$, historically reflect the idealistic structures, whereas the experimentally determined composition is nonstoichiometric and corresponds to fractional indexes. Boron-rich borides are usually characterized by large and complex unit cells, which can contain more than 1500 atomic sites and feature extended structures shaped as "tubes" and large modular polyhedra ("superpolyhedra"). Many of those sites have partial occupancy, meaning that the probability to find them occupied with a certain atom is smaller than one and thus that only some of them are filled with atoms. Scandium is distinguished among the rare-earth elements by that it forms numerous borides with uncommon structure types; this property of scandium is attributed to its relatively small atomic and ionic radii.

Crystals of the specific rare-earth boride YB_{66} are used as X-ray monochromators for selecting X-rays with certain energies (in the 1–2 keV range) out of synchrotron radiation. Other rare-earth borides may find application as thermoelectric materials, owing to their low thermal conductivity; the latter originates from their complex, "amorphous-like", crystal structure.

Metal Borides

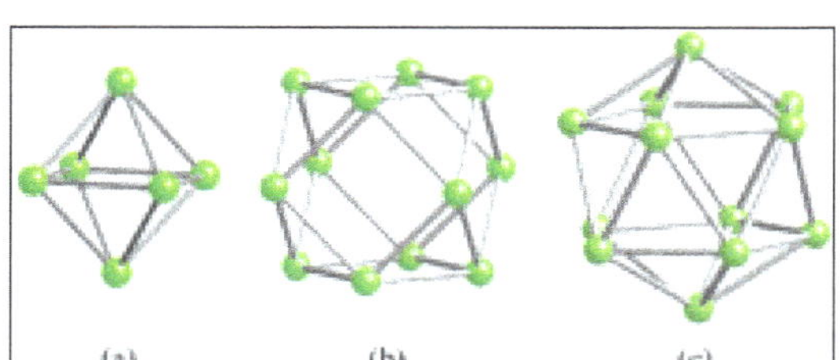

(a) B_6 octahedron, (b) B_{12} cuboctahedron and (c) B_{12} icosahedron.

In metal borides, the bonding of boron varies depending on the atomic ratio B/M. Diborides have B/M = 2, as in the well-known superconductor MgB_2; they crystallize in a hexagonal AlB_2-type layered structure. Hexaborides have B/M = 6 and form a three-dimensional boron framework based on a boron octahedron. Tetraborides, i.e. B/M = 4, are mixtures of diboride and hexaboride structures. Cuboctahedron is the structural unit of dodecaborides, which have a cubic lattice and B/M = 12. When the composition ratio exceeds 12, boron forms B_{12} icosahedra which are linked into a three-dimensional boron framework, and the metal atoms reside in the voids of this framework.

This complex bonding behavior originates from the fact that boron has only three valence electrons; this hinders tetrahedral bonding as in diamond or hexagonal bonding as in graphite. Instead, boron atoms form polyhedra. For example, three boron atoms make up a triangle where they share two electrons to complete the so-called three-center bonding. Boron polyhedra, such as B_6 octahedron, B_{12} cuboctahedron and B_{12} icosahedron, lack two valence electrons per polyhedron to complete the polyhedron-based framework structure. Metal atoms need to donate two electrons per boron polyhedron to form boron-rich metal borides. Thus, boron compounds are often regarded as electron-deficient solids.

Icosahedral B_{12} compounds include α-rhombohedral boron ($B_{13}C_2$), β-rhombohedral boron (MeB_x, $23 \leq x$), α-tetragonal boron ($B_{48}B_2C_2$), β-tetragonal boron (β-AlB_{12}), AlB_{10} or AlC_4B_{24}, YB_{25}, YB_{50}, YB_{66}, NaB_{15} or $MgAlB_{14}$, γ-AlB_{12}, BeB_3 and SiB_6.

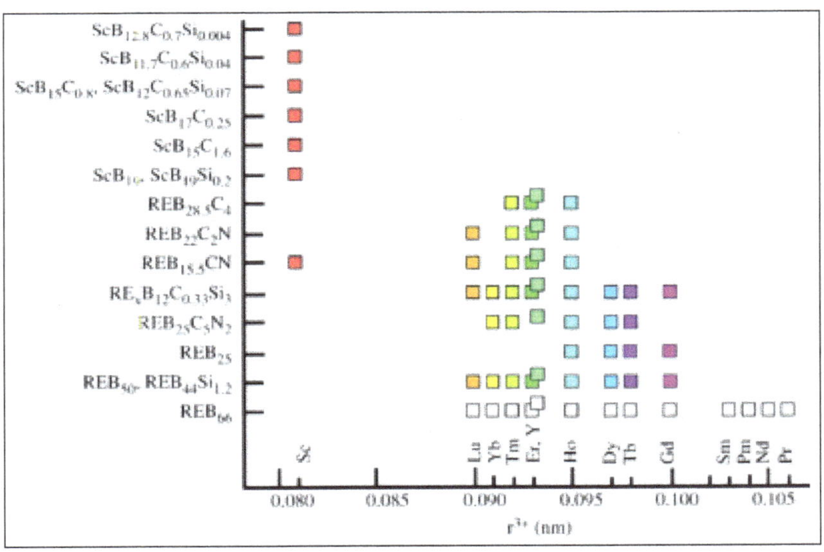

Relationship between the ionic radius of trivalent rare-earth ion and chemical composition of icosahedron-based rare-earth borides.

YB_{25} and YB_{50} decompose without melting that hinders their growth as single crystals by the floating zone method. However, addition of a small amount of Si solves this problem and results in single crystals with the stoichiometry of $YB_{41}Si_{1.2}$. This stabilization technique allowed the synthesis of some other boron-rich rare-earth borides.

Albert and Hillebrecht reviewed binary and selected ternary boron compounds containing main-group elements, namely, borides of the alkali and alkaline-earth metals, aluminum borides and compounds of boron and the nonmetals C, Si, Ge, N, P, As, O, S and Se. They, however, excluded the described here icosahedron-based rare-earth borides. Note that rare-earth elements have *d*- and *f*-electrons that complicates chemical and physical properties of their borides. Werheit *et al.* reviewed Raman spectra of numerous icosahedron-based boron compounds.

Figure shows a relationship between the ionic radius of trivalent rare-earth ions and the composition of some rare-earth borides. Note that scandium has many unique boron compounds, as shown in figure, because of the much smaller ionic radius compared with other rare-earth elements.

In understanding the crystal structures of rare-earth borides, it is important to keep in mind the concept of partial site occupancy, that is, some atoms in the described below unit cells can take several possible positions with a given statistical probability. Thus, with the given statistical probability, some of the partial-occupancy sites in such a unit cell are empty, and the remained sites are occupied.

$REAlB_{14}$ and REB_{25}

Compounds that were historically given the formulae $REAlB_{14}$ and REB_{25} have the $MgAlB_{14}$ structure with an orthorhombic symmetry and space group *Imma* (No. 74). In this structure, rare-earth atoms enter the Mg site. Aluminium sites are empty for REB_{25}. Both metal sites of $REAlB_{14}$ structure have partial occupancies of about 60–70%, which shows that the compounds are actually non-stoichiometric. The REB_{25} formula merely reflects the average atomic ratio [B]/[RE] = 25. Yttrium borides form both $YAlB_{14}$ and YB_{25} structures. Experiments have confirmed that the borides based on rare-earth elements from Tb to Lu can have the $REAlB_{14}$ structure. A subset of these borides, which contains rare-earth elements from Gd to Er, can also crystallize in the REB_{25} structure.

Korsukova *et al.* analyzed the $YAlB_{14}$ crystal structure using a single crystal grown by the high-temperature solution-growth method. The lattice constants were deduced as $a = 0.58212(3)$, $b = 1.04130(8)$ and $c = 0.81947(6)$ nm, and the atomic coordinates and site occupancies.

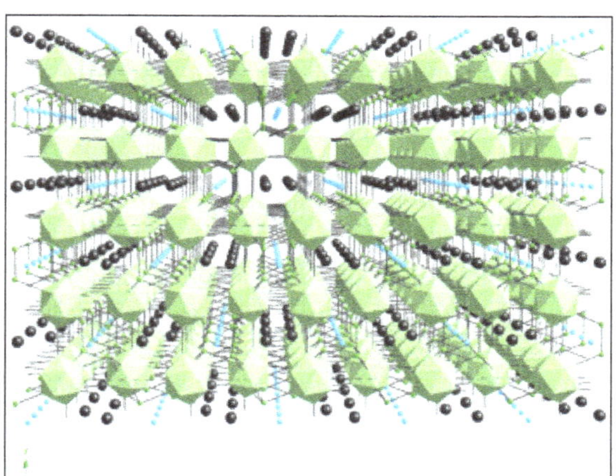

Crystal structure of $YAlB_{14}$. Black and blue spheres indicate Y and Al atoms, respectively. Vacancies at the Y and Al sites are ignored.

Figure shows the crystal structure of $YAlB_{14}$ viewed along the *x*-axis. The large black spheres are Y atoms, the small blue spheres are Al atoms and the small green spheres are the bridging boron sites; B_{12} clusters are depicted as the green icosahedra. Boron framework of $YAlB_{14}$ is one of the simplest among icosahedron-based borides – it consists of only one kind of icosahedra and one bridging boron site. The bridging boron site is tetrahedrally coordinated by four boron atoms. Those atoms are another boron atom in the counter bridge site and three equatorial boron atoms of one of three B_{12} icosahedra. Aluminium atoms are separated by 0.2911 nm and are arranged in lines parallel to the *x*-axis, whereas yttrium atoms are separated by 0.3405 nm. Both the Y atoms and B_{12} icosahedra form zigzags along the *x*-axis. The bridging boron atoms connect three equatorial boron atoms of three icosahedra and those icosahedra make up a network parallel to the (101) crystal plane (*x-z* plane in the figure). The bonding distance between the bridging boron and the equatorial boron atoms is 0.1755 nm, which is typical for the strong covalent B-B bond (bond length 0.17–0.18 nm); thus, the bridging boron atoms strengthen the individual network planes. On the other hand, the large distance between the boron atoms within the bridge (0.2041 nm) suggests weaker interaction, and thus the bridging sites contribute little to the bonding between the network planes.

The boron framework of $YAlB_{14}$ needs donation of four electrons from metal elements: two electrons for a B_{12} icosahedron and one electron for each of the two bridging boron atoms – to support their tetrahedral coordination. The actual chemical composition of $YAlB_{14}$, determined by the structure analysis, is $Y_{0.62}Al_{0.71}B_{14}$. If both metal elements are trivalent ions then 3.99 electrons can be transferred to the boron framework, which is very close to the required value of 4. However, because the bonding between the bridging boron atoms is weaker than in a typical B-B covalent bond, less than 2 electrons are donated to this bond, and metal atoms need not be trivalent. On the other hand, the electron transfer from metal atoms to the boron framework implies that not only strong covalent B-B bonding within the framework but also ionic interaction between metal atoms and the framework contribute to the $YAlB_{14}$ phase stabilization.

REB_{66}-type Borides

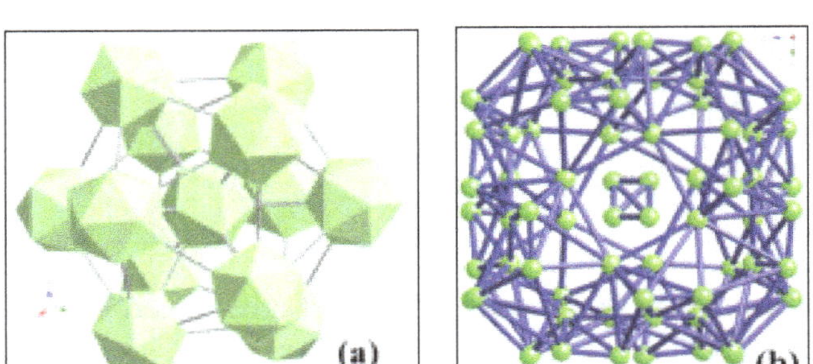

(a) Thirteen-icosahedron unit $(B_{12})_{13}B_{12}$ (supericosahedron), and (b) B_{80} cluster unit. The excessive bonding in panel (b) is because it assumes that all sites are occupied, whereas the total number of boron atoms is only 42.

In addition to yttrium, a wide range of rare-earth elements from Nd to Lu, except for Eu, can form REB_{66} compounds. Seybolt discovered the compound YB_{66} in 1960 and its structure was solved by Richards and Kasper in 1969. They reported that YB_{66} has a face-centered cubic structure with space group *Fm3c* (No. 226) and lattice constant a = 2.3440(6) nm. There are 13 boron sites B1–B13 and one yttrium site. The B1 sites form one icosahedron and the B2–B9 sites make up another icosahedron. These icosahedra arrange in a thirteen-icosahedron unit $(B_{12})_{12}B_{12}$ which is shown in figure and is called supericosahedron. The icosahedron formed by the B1 site atoms is located at the center of the supericosahedron. The supericosahedron is one of the basic units of the boron framework of YB_{66}. There are two types of supericosahedra: one occupies the cubic face centers and another, which is rotated by 90°, is located at the center of the cell and at the cell edges. Thus, there are eight supericosahedra (1248 boron atoms) in the unit cell.

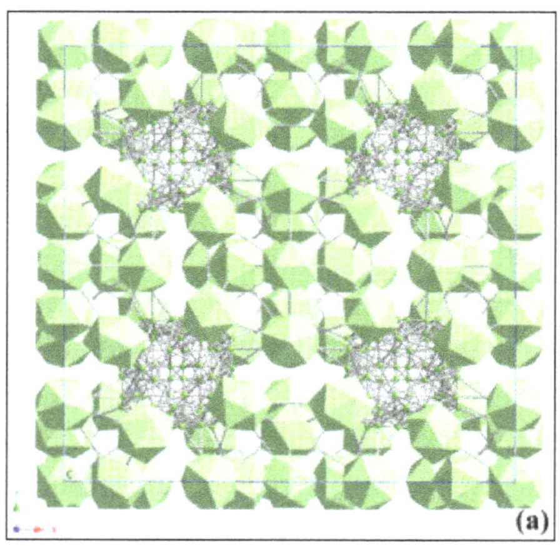

The boron framework of YB_{66} viewed along the z-axis.

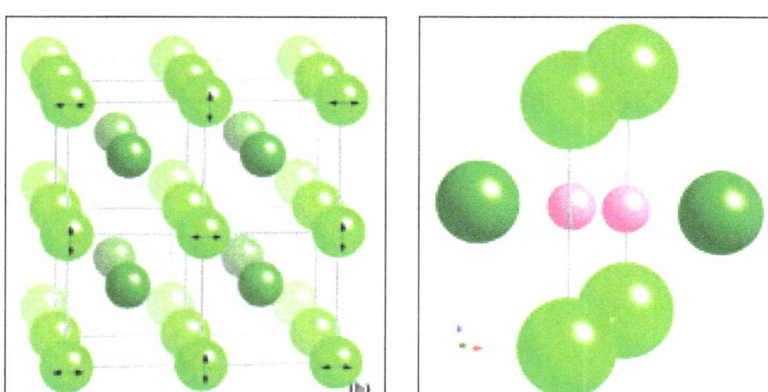

Schematically drawn boron framework of YB_{66}. Light green spheres show the boron supericosahedra and their relative orientations are indicated by arrows. Dark green spheres correspond to the B_{80} clusters. Pair of Y sites (pink spheres) in YB_{66}. Light green spheres show the boron supericosahedron and dark green spheres correspond to the B_{80} clusters.

Number of B, Y and Mo or Pt atoms in the unit cell deduced from chemical composition, lattice constant and density (experiment).					
Composition	a(nm)	$\rho(g/cm^3)$	N_B	N_y	$N_{Mo/Pt}$
YB_{66}	2.3440	2.52	1610	24.4	–
$YB_{61.754}$	2.3445	2.5687	1628	26.4	–
YB_{62}	2.34364	2.5662	1624	26.2	–
YB_{56}	2.34600	2.5927	1626	29.0	–
$YMo_{0.20}B_{62.4}$	2.34258	2.64	1628	26.1	5.3
$YPt_{0.091}B_{63.5}$	2.34300	2.6344	1634	25.7	2.4
$YPt_{0.096}B_{63.3}$	2.34223	2.6355	1630	25.7	2.5
$YPt_{0.14}B_{62.0}$	2.34055	2.6762	1629	26.3	3.7

Another structure unit of YB_{66}, shown in figure, is B_{80} cluster of 80 boron sites formed by the B10 to B13 sites. All those 80 sites are partially occupied and in total contain only about 42 boron atoms. The B_{80} cluster is located at the body center of the octant of the unit cell, i.e., at the 8*a* position (1/4, 1/4, 1/4); thus, there are eight such clusters (336 boron atoms) per unit cell. Two independent structure analyses came to the same conclusion that the total number of boron atoms in the unit cell is 1584. The boron framework structure of YB_{66} is shown in figure. To indicate relative orientations of the supericosahedra, a schematic drawing is shown in figure, where the supericosahedra and the B_{80} clusters are depicted by light green and dark green spheres, respectively; at the top surface of the unit cell, the relative orientations of the supericosahedra are indicated by arrows. There are 48 yttrium sites ((0.0563, 1/4, 1/4) for YB_{62}) in the unit cell. Richards and Kasper fixed the Y site occupancy to 0.5 that resulted in 24 Y atoms in the unit cell and the chemical composition of YB_{66}. As shown in figure, Y sites form a pair separated by only 0.264 nm in YB_{62}. This pair is aligned normal to the plane formed by four supericosahedra. The Y site occupancy 0.5 implies that the pair has always one Y atom with one empty site.

x-coordinates and occupancies of the Y1 and Y2 sites.		
Atom	x	Occupancy
Y1	0.0542(3)	0.437(9)
Y2	0.0725(11)	0.110(12)

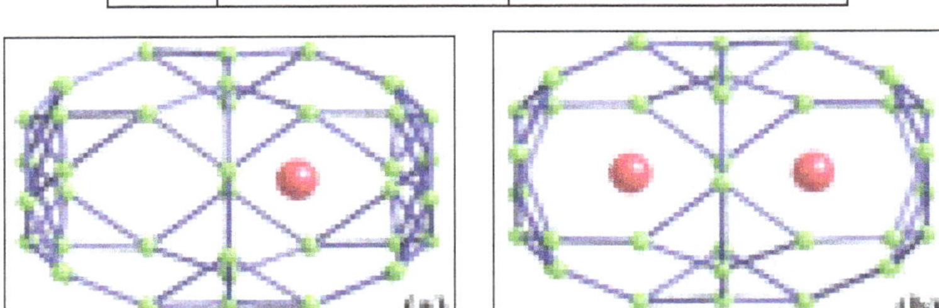

Two cases of the Y pair sites: with one Y atom (top) and two Y atoms (bottom). In the latter case, some neighboring boron sites are eliminated because they are too close to the Y site.

Slack *et al.* reported that the total number of boron atoms in the unit cell, calculated from the measured values of density, chemical composition and lattice constant, is 1628 ± 4, which is larger than the value 1584 obtained from the structural analysis. The number of B atoms in the unit cell remains nearly constant when the chemical composition changes from YB_{56} to YB_{66}. On the other hand, the total number of yttrium atoms per unit cell varies, and it is, for example, ~26.3 for YB_{62} . If the total number of Y atoms stays below or equal to 24 then it is possible that one Y atom accommodates in each Y pair (partial occupancy). However, the experimental value of 26.3 significantly exceeds 24, and thus both pair sites might be occupied. In this case, because of the small separation between the two Y atoms, they must be repelled by the Coulomb force. To clarify this point, split Y sites were introduced in the structure analysis resulting in a better agreement with the experiment. The Y site distances and occupancies are presented in the left table.

There are twenty Y pair sites with one Y atom and three pairs with two Y atoms; there is also one empty Y pair (partial occupancy = 0). The separation 0.340 nm for the Y2 pair site (two Y atoms in the pair site) is much larger than the separation 0.254 nm for the Y1 pair site (one Y atom in the pair site), as expected. The total number of Y atoms in the unit cell is 26.3, exactly as measured. Both cases are compared in figure. The larger separation for the Y2 pair site is clear as compared with that for the Y1 pair site. In case of the Y2 pair, some neighboring boron sites that belong to the B_{80} cluster must be unoccupied because they are too close to the Y2 site.

Splitting the Y site yields right number of Y atoms in the unit cell, but not B atoms. Not only the occupation of the B sites in the B_{80} cluster must be strongly dependent on whether or not the Y site is the Y1 state or the Y2 state, but also the position of the occupied B sites must be affected by the state of the Y site. Atomic coordinates and site occupancies are summarized in table.

$REB_{41}Si_{1.2}$

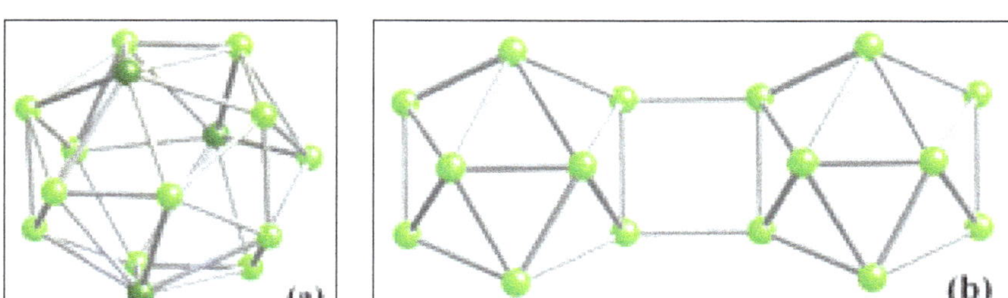

(a) $B_{12}Si_3$ polyhedron unit. Darker green spheres represent the sites which are occupied either by Si or B atoms. (b) Unusual linkage between the B_{12}-I5 icosahedra connected via two apex atoms of each icosahedron.

Similar to yttrium, rare-earth metals from Gd to Lu can form $REB_{41}Si_{1.2}$-type boride. The first such compound was synthesized by solid-state reaction and its structure was deduced as YB_{50}. X-ray powder diffraction (XRD) and electron diffraction indicated that

YB_{50} has an orthorhombic structure with lattice constants a = 1.66251(9), b = 1.76198 and c = 0.94797(3) nm. The space group was assigned as $P2_12_12$. Because of the close similarity in lattice constants and space group, one might expect that YB_{50} has the γ-AlB_{12}-type orthorhombic structure whose lattice constants and space group are a = 1.6573(4), b = 1.7510(3) and c = 1.0144(1) nm and $P2_12_12$. YB_{50} decomposes at ~1750 °C without melting that hinders growth of single crystals from the melt. Small addition of silicon made YB_{50} to melt without decomposition, and so enabled single-crystal growth from the melt and single-crystal structure analysis.

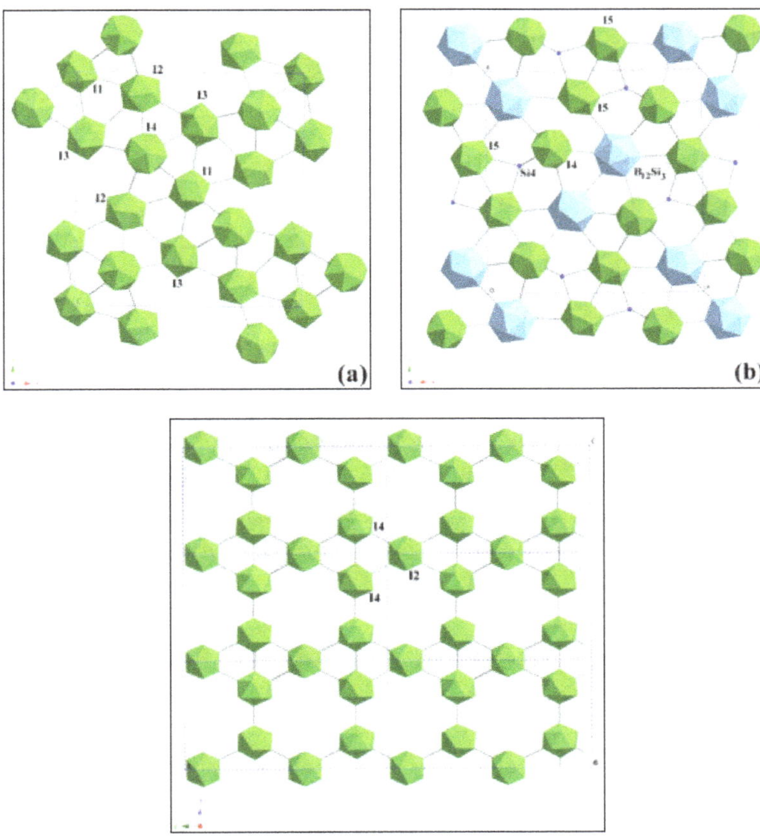

(a) The boron network that consists of the icosahedra I1, I2 and I3 and is located at z = 0 plane. The icosahedron I4 lies above and below of this network at z = ± 0.25. (b) The boron network that consists of the I5 icosahedron and $B_{12}Si_3$ polyhedron (blue) and is located at z = 0.5 plane. The icosahedron I4 lies above and below this network at z = 0.25 and 0.75. The b-c network formed by icosahedra I4 and I2 as seen along the a-axis. The network is drawn within the range x = 0.09–0.41.

The structure analysis indicated that $YB_{41}Si_{1.2}$ has not the γ-AlB_{12}-type lattice but a rare orthorhombic crystal structure (space group: *Pbam*, No. 55) with lattice constants of a = 1.674(1) nm, b = 1.7667(1) nm and c = 0.9511(7) nm. There are 58 independent atomic sites in the unit cell. Three of them are occupied by either B or Si atoms (mixed-occupancy sites), one is a Si bridge site and one is Y site. From the remaining 53 boron sites, 48 form icosahedra and 5 are bridging sites.

The boron framework of $YB_{41}Si_{1.2}$ consists of five B_{12} icosahedra (I1–I5) and a $B_{12}Si_3$ polyhedron shown in figure. An unusual linkage is depicted in figure, where two B_{12}-I5 icosahedra connect via two B atoms of each icosahedron forming an imperfect square. The boron framework of $YB_{41}Si_{1.2}$ can be described as a layered structure where two boron networks stack along the *z*-axis. One boron network consists of 3 icosahedra I1, I2 and I3 and is located in the $z = 0$ plane; another network consists of the icosahedron I5 and the $B_{12}Si_3$ polyhedron and lies at $z = 0.5$. The icosahedron I4 bridges these networks, and thus its height along the *z*-axis is 0.25.

The I4 icosahedra link two networks along the *c*-axis and therefore form an infinite chain of icosahedra along this axis as shown in figure. The unusually short distances (0.4733 and 0.4788 nm) between the neighboring icosahedra in this direction result in the relatively small *c*-axis lattice constant of 0.95110(7) nm in this compound – other borides with a similar icosahedral chain have this value larger than 1.0 nm. However, the bonding distances between the apex B atoms (0.1619 and 0.1674 nm) of neighboring I4 icosahedra are usual for the considered metal borides.

Another unusual feature of $YB_{41}Si_{1.2}$ is the 100% occupancy of the Y site. In most icosahedron-based metal borides, metal sites have rather low site occupancy, for example, about 50% for YB_{66} and 60–70% for $REAlB_{14}$. When the Y site is replaced by rare-earth elements, $REB_{41}Si_{1.2}$ can have an antiferromagnetic-like ordering because of this high site occupancy.

Homologous Icosahedron-based Rare-earth Borides

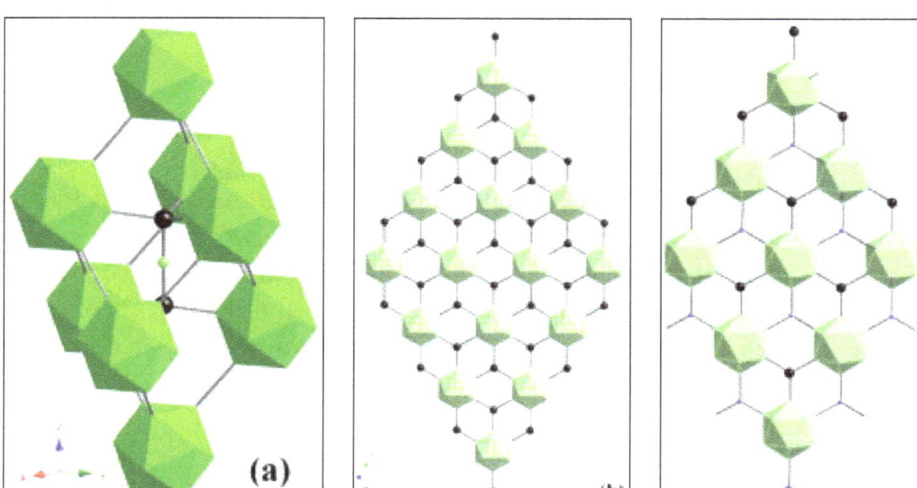

(a) Structure unit of B_4C and (b) *c*-plane network of B_{12} icosahedra in the B_4C structure. B_{12} icosahedron network bridged by nitrogen (blue) and carbon (black) atoms.

Rare-earth borides $REB_{15.5}CN$, $REB_{22}C_2N$ and $REB_{28.5}C_4$ are homologous, i.e. have a similar crystal structure, to B_4C. The latter has a structure typical of icosahedron-based borides, as shown in figure. There, B_{12} icosahedra form a rhombohedral lattice unit (space group: $R3m$ (No. 166), lattice constants: a = 0.56 nm and c = 1.212 nm) surrounding a C-B-C chain that resides at the center of the lattice unit, and both C atoms bridge the neighboring three icosahedra. This structure is layered: as shown in figure, B_{12} icosahedra and bridging carbons form a network plane that spreads parallel to the c-plane and stacks along the c-axis.

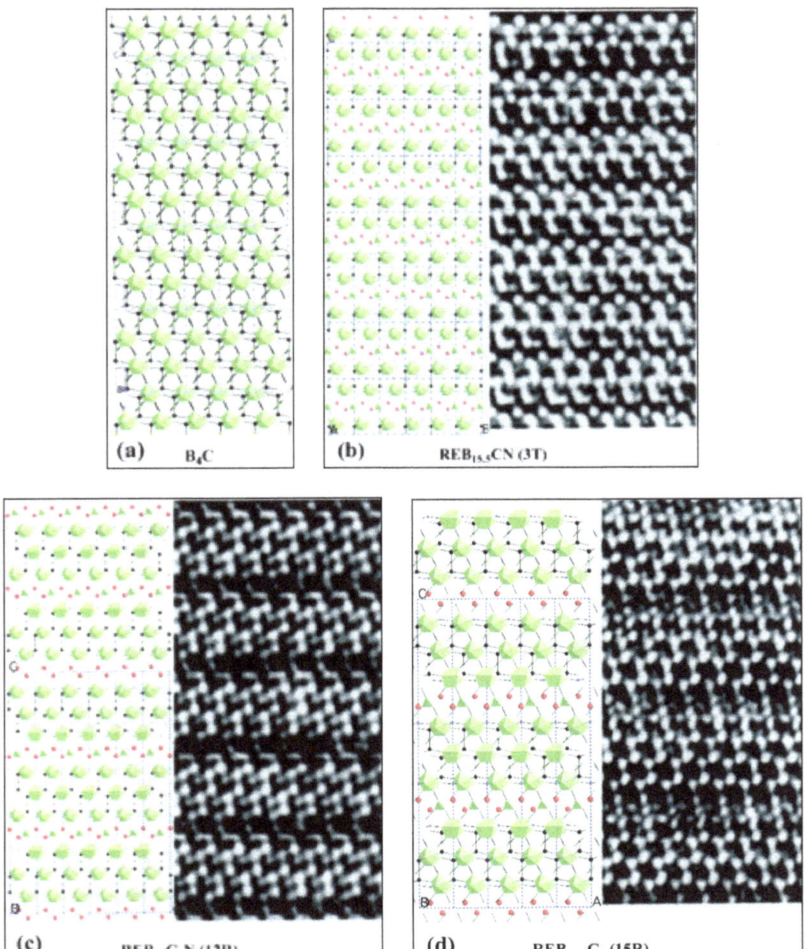

Stacking sequences of homologous icosahedron-based rare-earth borides and their HR-TEM lattice images; (a) B_4C, (b) $REB_{15.5}CN$ (3T), (c) $REB_{22}C_2N$ (12R) and (d) $REB_{28.5}C_4$ (15R). Red circles are rare-earth atoms. HRTEM lattice images were obtained for $YB_{15.5}CN$, $YB_{22}C_2N$ and $YB_{28.5}C_4$ compounds.

These homologous compounds have two basic structure units – the B_{12} icosahedron and the B_6 octahedron. The network plane of B_4C structure can be periodically replaced by a B_6 octahedron layer so that replacement of every third, fourth and fifth layer would correspond to $REB_{15.5}CN$, $REB_{22}C_2N$ and $REB_{28.5}C_4$, respectively. The B_6 octahedron is smaller than the B_{12} icosahedron; therefore, rare-earth elements can reside in the space created by the replacement. The stacking sequences of B_4C, $REB_{15.5}CN$, $REB_{22}C_2N$ and $REB_{28.5}C_4$ are shown in figures above a, b, c and d, respectively. High-resolution transmission electron microscopy (HRTEM) lattice images of the latter three compounds, added to figure., do confirm the stacking sequence of each compound. The symbols 3T, 12R and 15R in brackets indicate the number of layers necessary to complete the stacking sequence, and T and R refer to trigonal and rhombohedral. Thus, $REB_{22}C_2N$ and $REB_{28.5}C_4$ have rather large *c*-lattice constants.

Because of the small size of the B_6 octahedra, they cannot interconnect. Instead, they bond to the B_{12} icosahedra in the neighboring layer, and this decreases bonding strength in the *c*-plane. Nitrogen atoms strengthen the bonding in the *c*-plane by bridging three icosahedra, like C atoms in the C-B-C chain. Figure depicts the *c*-plane network revealing the alternate bridging of the boron icosahedra by N and C atoms. Decreasing the number of the B_6 octahedra diminishes the role of nitrogen because the C-B-C chains start bridging the icosahedra. On the other hand, in MgB_9N the B_6 octahedron layer and the B_{12} icosahedron layer stack alternatively and there is no C-B-C chains; thus only N atoms bridge the B_{12} icosahedra. However, REB_9N compounds have not been identified yet.

Sc, Y, Ho, Er, Tm and Lu are confirmed to form $REB_{15.5}CN$-type compounds. Single-crystal structure analysis yielded trigonal symmetry for $ScB_{15.5}CN$ (space group *P3m1* (No.164) with a = 0.5568(2) and c = 1.0756(2) nm), and the deduced atomic coordinates.

$REB_{22}C_2N$ was synthesized for Y, Ho, Er, Tm and Lu. The crystal structure, solved for a representative compound $YB_{22}C_2N$, belongs to the trigonal with space group *R3m* (No.166); it has six formula units in the unit cell and lattice constants $a = b$ = 0.5623(0) nm and c = 4.4785(3) nm. Atomic coordinates of $YB_{22}C_2N$.

Y, Ho, Er, Tm and Lu also form $REB_{28.5}C_4$ which has a trigonal crystal structure with space group *R3m* (No. 166). Lattice constants of the representative compound $YB_{28.5}C_4$ are $a = b$ = 0.56457(9) nm and c = 5.68873(13) nm and there are six formula units in the unit cell. Structure data of $YB_{28.5}C_4$.

$RE_xB_{12}C_{0.33}Si_{3.0}$

Initially these were described as ternary RE-B-Si compounds, but later carbon was included to improve the structure description that resulted in a quaternary RE-B-C-Si composition. $RE_xB_{12}C_{0.33}Si_{3.0}$ (RE=Y and Gd–Lu) have a unique crystal structure with two units – a cluster of B_{12} icosahedra and a Si_8 ethane-like complex – and one bonding configuration $(B_{12})_3$≡Si-C≡$(B_{12})_3$. A representative compound of this group is $Y_xB_{12}C_{0.33}Si_{3.0}$ (x=0.68).

It has a trigonal crystal structure with space group $R\bar{3}m$ (No. 166) and lattice constants $a = b = 1.00841(4)$ nm, $c = 1.64714(5)$ nm, $\alpha = \beta = 90°$ and $\gamma = 120°$.

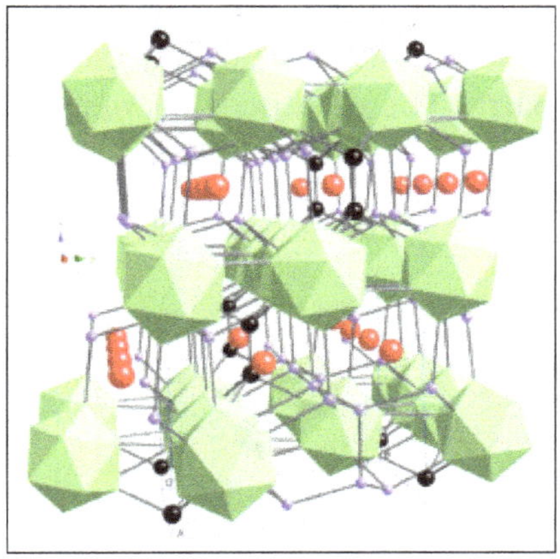

Crystal structure of $RE_xB_{12}C_{0.33}Si_{3.0}$ (RE=Y or Dy) viewed along the direction close to. Red, black and blue spheres correspond to Y/Dy, C and Si atoms, respectively. Vacancies at the Y/Dy site are ignored.

The crystal has layered structure. Figure shows a network of boron icosahedra that spreads parallel to the (001) plane, connecting with four neighbors through B1–B1 bonds. The C3 and Si3 site atoms strengthen the network by bridging the boron icosahedra. Contrary to other boron-rich icosahedral compounds, the boron icosahedra from different layers are not directly bonded. The icosahedra within one layer are linked through Si_8 ethane-like clusters with $(B_{12})_3{\equiv}Si\text{-}C{\equiv}(B_{12})_3$ bonds, as shown in figures 16a and b.

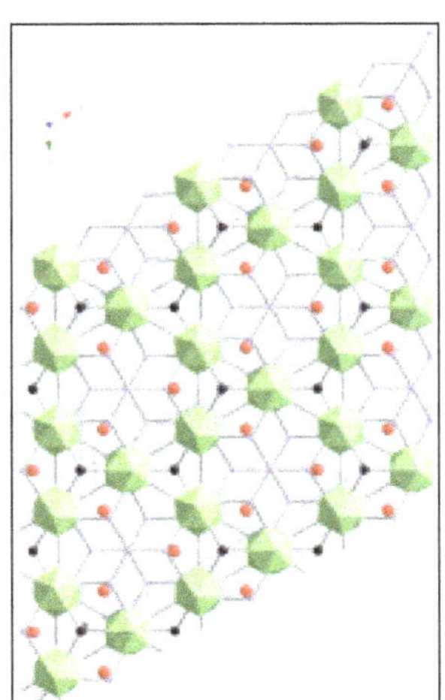

A network of boron icosahedra lying in the (001) plane. Black, blue and red spheres correspond to C, Si and Y atoms, respectively.

There are eight atomic sites in the unit cell: one yttrium Y, four boron B1–B4, one carbon C3 and three silicon sites Si1–Si3. Atomic coordinates, site occupancy and isotropic displacement factors are listed in table Va; 68% of the Y sites are randomly occupied and remaining Y sites are vacant. All boron sites and Si1 and Si2 sites are fully occupied. The C3 and Si3 sites can be occupied by either carbon or silicon atoms (mixed occupancy) with a probability of about 50%. Their separation is only 0.413 Å, and thus either the C3 or Si3 sites, but not both, are occupied. These sites form Si-C pairs, but not Si-Si or C-C pairs. The distances between the C3 and Si3 sites and the surrounding sites for $Y_xB_{12}C_{0.33}Si_{3.0}$ are summarized in table Vb and the overall crystal structure is shown in figure.

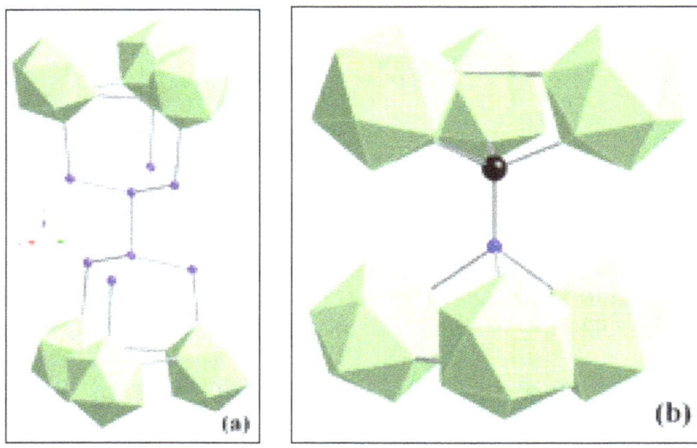

(a) Si_8 ethane-like cluster connects layers of boron icosahedra; a layer of boron icosahedra that lies at the same level as the Si_8 cluster is not shown. (b) bonding configuration $(B_{12})_3$≡Si-C≡$(B_{12})_3$.

Salvador *et al.* reported an isotypic terbium compound $Tb_{3-x}C_2Si_8(B_{12})_3$. Most parts of the crystal structure are the same as those described above; however, its bonding configuration is deduced as $(B_{12})_3$≡C-C≡$(B_{12})_3$ instead of $(B_{12})_3$≡Si-C≡$(B_{12})_3$. The authors intentionally added carbon to grow single crystals whereas the previous crystals were accidentally contaminated by carbon during their growth. Thus, higher carbon concentration was achieved. Existence of both bonding schemes of $(B_{12})_3$≡Si-C≡$(B_{12})_3$ and $(B_{12})_3$≡C-C≡$(B_{12})_3$ suggests the occpancy of the carbon sites of 50–100%. On the other hand, $(B_{12})_3$≡Si-Si≡$(B_{12})_3$ bonding scheme is unlikely because of too short Si-Si distance, suggesting that the minimum carbon occupancy at the site is 50%. Some B atoms may replace C atoms at the C3 site, as previously assigned to the B site. However, the carbon occupation is more likely because the site is tetrahedrally coordinated whereas the B occupation of the site needs an extra electron to complete tetrahedral bonding. Thus, carbon is indispensable for this group of compounds.

Scandium Compounds

Scandium has the smallest atomic and ionic (3+) radii (1.62 and 0.885 Å, respectively) among the rare-earth elements. It forms several icosahedron-based borides which are not found for other rare-earth elements; however, most of them are ternary Sc-B-C compounds. There are many boron-rich phases in the boron-rich corner of Sc-B-C phase diagram, as shown in figure. A slight variation of the composition can produce

ScB_{19}, $ScB_{17}C_{0.25}$, $ScB_{15}C_{0.8}$ and $ScB_{15}C_{1.6}$; their crystal structures are unusual for borides and are very different from each other.

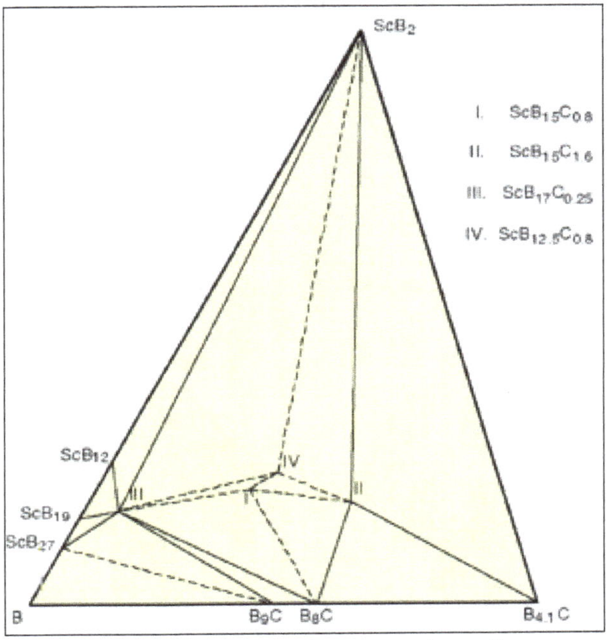

The boron-rich corner of the Sc-B-C phase diagram.

$ScB_{19+x}Si_y$

$ScB_{19+x}Si_y$ has a tetragonal crystal structure with space group $P4_12_12$ (No. 92) or $P4_32_12$ and lattice constants of a, b = 1.03081(2) and c = 1.42589(3) nm; it is isotypic to the α-AlB_{12} structure type. There are 28 atomic sites in the unit cell, which are assigned to 3 scandium atoms, 24 boron atoms and one silicon atom. Atomic coordinates, site occupancies and isotropic displacement factors.

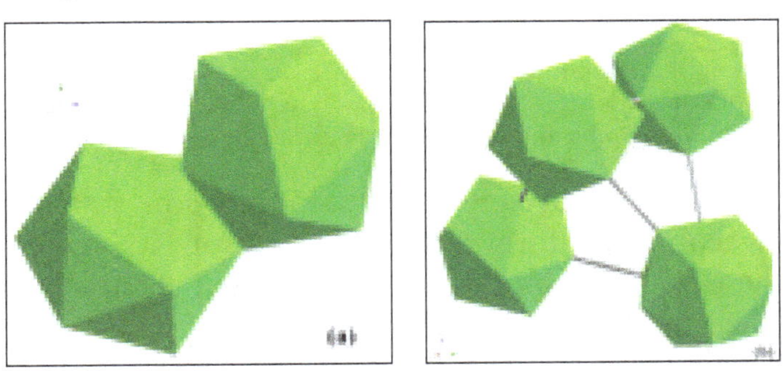
(a) Twinned B_{22} icosahedra, (b) boron supertetrahedron constructed by 4 icosahedral.

The boron framework of $ScB_{19+x}Si_y$ is based on one B_{12} icosahedron and one B_{22} unit. This unit can be observed in β-tetragonal boron and is a modification of the B_{20} unit of α-AlB_{12} (or B_{19} unit in early reports). The B_{20} unit is a twinned icosahedron made from B13 to B22 sites with two vacant sites and one B atom (B23) bridging both sides of the unit. The twinned icosahedron is shown in figure. B23 was treated as an isolated atom in the early reports; it is bonded to each twinned icosahedra through B18 and to another icosahedron through B5 site. If the twinned icosahedra were independent without twinning then B23 would be a bridge site linking three icosahedra. However, because of twinning, B23 shifts closer to the twinned icosahedra than another icosahedron; thus B23 is currently treated as a member of the twinned icosahedra. In $ScB_{19+x}Si_y$, the two B24 sites which correspond to the vacant sites in the B_{20} unit are partially occupied; thus, the unit should be referred to as a B_{22} cluster which is occupied by about 20.6 boron atoms. Scandium atoms occupy 3 of 5 Al sites of α-AlB_{12}, that is Sc1, Sc2 and Sc3 correspond to Al4, Al1 and Al2 sites of α-AlB_{12}, respectively. The Al3 and Al5 sites are empty for $ScB_{19+x}Si_y$, and the Si site links two B_{22} units. This phase also exists without silicon.

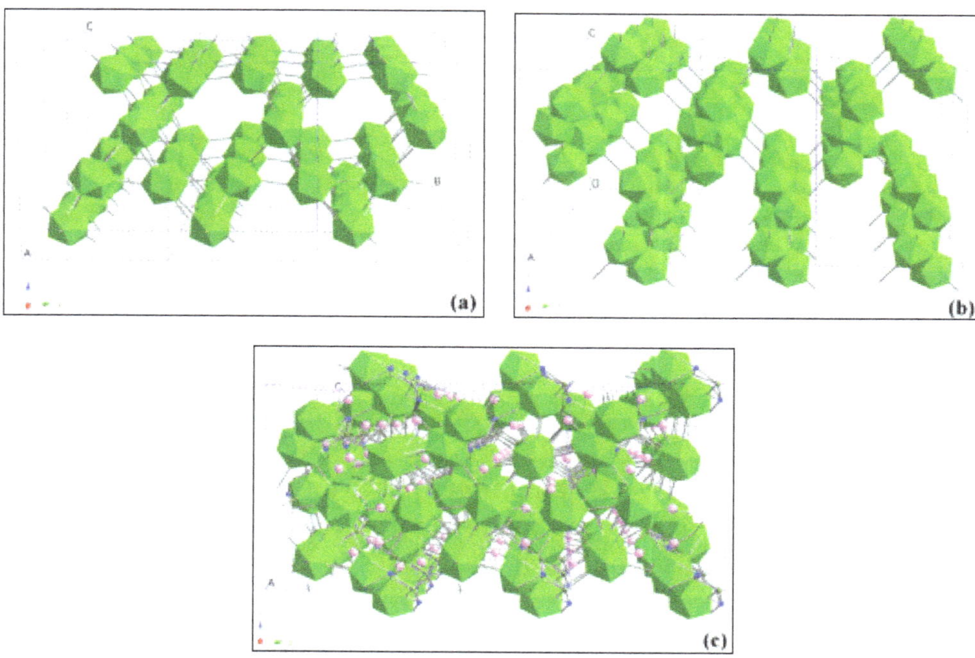

(a) Network of boron icosahedra, (b) B_{22} unit network and (c) overall crystal structure of $ScB_{19+x}Si_y$; pink and blue spheres indicate Sc and Si atoms, respectively.

Figure shows the network of boron icosahedra in the boron framework of $ScB_{19+x}Si_y$. In this network, 4 icosahedra form a supertetrahedron; its one edge is parallel to the *a*-axis, and the icosahedra on this edge make up a chain along the *a*-axis. The opposite edge of the supertetrahedron is parallel to the *b*-axis and the icosahedra on this edge form a chain along the *b*-axis. As shown in figure, there are wide tunnels surrounded by the icosahedron arrangement along the *a*- and *b*-axes. The tunnels are filled by the B_{22} units which strongly bond to the surrounding icosahedra; the connection of the B_{22} units is helical and it runs along the *c*-axis as shown in figure. Scandium atoms occupy the voids in the boron network as shown in figure, and the Si atoms bridge the B_{22} units.

$ScB_{17}C_{0.25}$

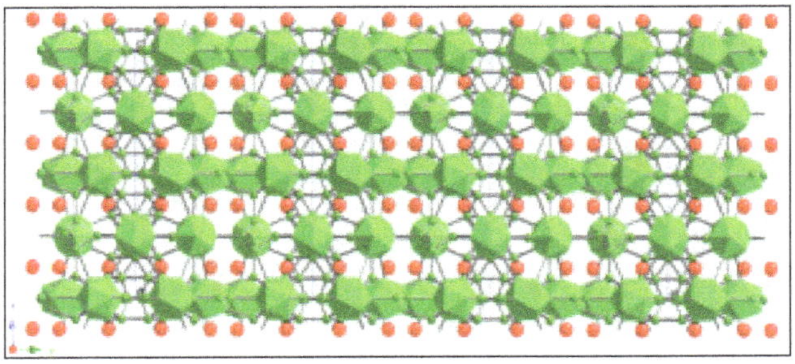

$ScB_{17}C_{0.25}$ crystal structure viewed along the *a*-axis. The icosahedron layers alternatively stack along the *c*-axis in the order I1–I2–I1–I2–I1.

Very small amount of carbon is sufficient to stabilize "$ScB_{17}C_{0.25}$". This compound has a broad composition range, namely $ScB_{16.5+x}C_{0.2+y}$ with $x \leq 2.2$ and $y \leq 0.44$. $ScB_{17}C_{0.25}$ has a hexagonal crystal structure with space group *P6mmm* (No. 199) and lattice constants $a, b = 1.45501(15)$ nm and $c = 0.84543(16)$ nm.

There are 19 atomic sites in the unit cell, which are assigned to one scandium site Sc, 14 boron sites B1–B14 having 100% occupancy, two boron-carbon mixed-occupancy sites B/C15 and B/C16, and two partial-occupancy boron sites B17 and B18. Atomic coordinates, site occupancies and isotropic displacement factors are listed in table. Although a very small amount of carbon (less than 2 wt%!) plays an important role in the phase stability, carbon does not have its own sites but shares with boron two interstitial sites B/C15 and B/C16.

There are two inequivalent B_{12} icosahedra, I1 and I2, which are constructed by the B1–B5 and B8–B12 sites, respectively. A "tube" is another characteristic structure unit of $ScB_{17}C_{0.25}$. It extends along the *c*-axis and consists of B13, B14, B17 and B18 sites where B13 and B14 form 6-membered rings. B17 and B18 sites also form 6-membered rings; however, their mutual distances (0.985 Å for B17 and 0.955 Å for B18) are too short for a simultaneous occupation of the neighboring sites. Therefore, boron atoms occupy 2nd neighbor site forming a triangle. The occupancies of B17 and B18 sites should be 50%, but the structure analysis suggests larger values. The crystal structure viewed along the *a*-axis is shown in figure, which suggests that the $ScB_{17}C_{0.25}$ is a layered material. Two layers, respectively constructed by the icosahedra I1 and I2, alternatively stack along the *c*-axis. However, the $ScB_{17}C_{0.25}$ crystal is not layered.

For example, during arc-melting, $ScB_{17}C_{0.25}$ needle crystals violently grow along the *c*-axis – this never happens in layered compounds. The crystal structure viewed along the *c*-axis is shown in figure. The icosahedra I1 and I2 form a ring centered by the "tube" shown in figure, which probably governs the properties of the $ScB_{17}C_{0.25}$ crystal. B/C15 and B/C16 mixed-occupancy sites interconnect the rings. A structural similarity can be seen between $ScB_{17}C_{0.25}$ and BeB_3.

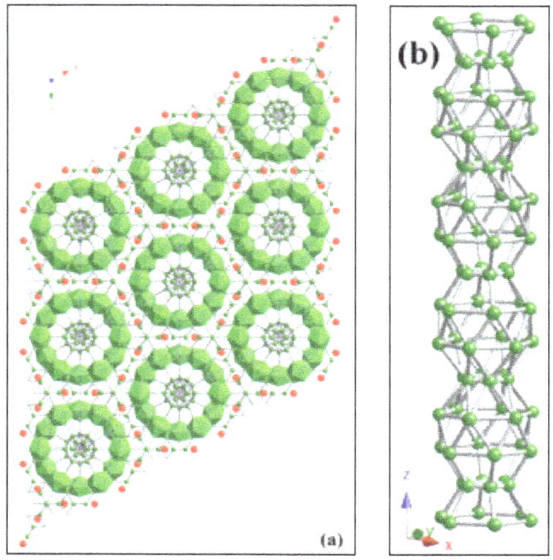

(a) $ScB_{17}C_{0.25}$ crystal structure viewed along the *c*-axis. Icosahedra I1 and I2 form a ring centered by a "tube". (b) The "tube" structure that runs along the *c*-axis. Partial occupancies of B17 and B18 are ignored.

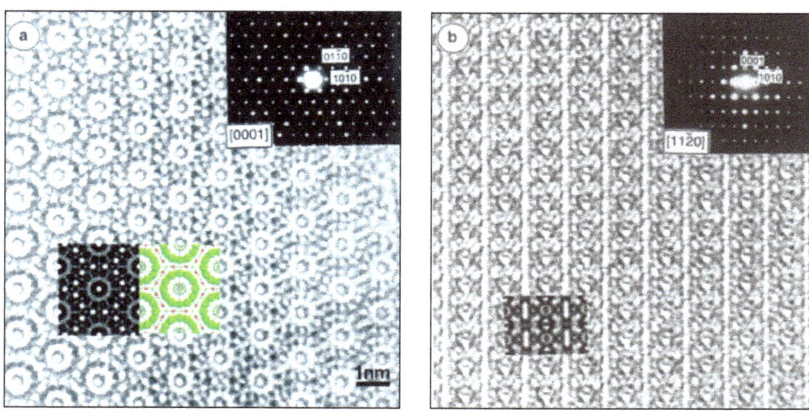

Experimental HRTEM lattice images and electron diffraction patterns (top-right insets) taken along the (a) and (b) directions. Image simulations are added in the bottom-left insets, and a fragment of the crystal structure is also added to (a).

Figures present HRTEM lattice images and electron diffraction patterns taken along the and crystalline directions, respectively. The HRTEM lattice image of figure a reproduces well the (*a, b*) plane of the crystal structure shown in figure, with the clearly visible rings membered by icosahedra I1 and I2 and centered by the "tube". Figure proves that $ScB_{17}C_{0.25}$ does not have layered character but its *c*-axis direction is built up by the ring-like structure and tubular structures.

$Sc_{0.83-x}B_{10.0-y}C_{0.17+y}Si_{0.083-z}$

$Sc_{0.83-x}B_{10.0-y}C_{0.17+y}Si_{0.083-z}$ ($x = 0.030$, $y = 0.36$ and $z = 0.026$) has a cubic crystal structure with space group $F43m$ (No. 216) and lattice constant $a = 2.03085(5)$ nm. This compound was initially identified as $ScB_{15}C_{0.8}$ (phase I in the Sc-B-C phase). A small amount of Si was added into the floating zone crystal growth and thus this phase is a quaternary compound. Its rare cubic structure has 26 sites in the unit cell: three Sc sites, two Si sites, one C site and 20 B sites; 4 out of 20 B sites are boron-carbon mixed-occupancy sites. Atomic coordinates, site occupancies and isotropic displacement factors.

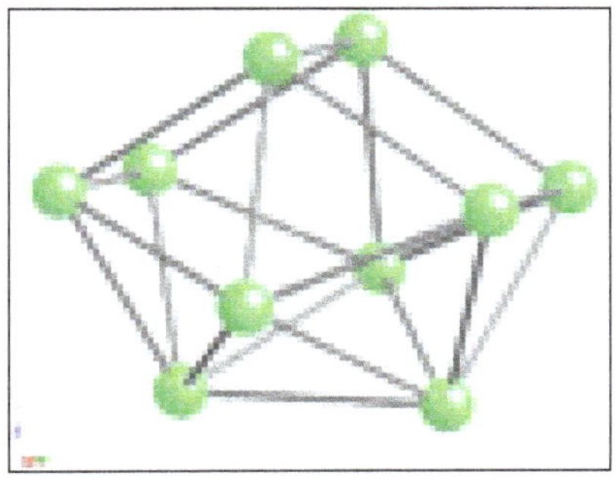

B_{10} polyhedron in the $Sc_{0.83-x}B_{10.0-y}C_{0.17+y}Si_{0.083-z}$ crystal structure.

In the unit cell, there are three independent icosahedra, I1, I2 and I3, and a B_{10} polyhedron which are formed by the B1–B4, B5–B8, B9–B13 and B14–B17 sites, respectively. The B_{10} polyhedron has not been observed previously and it is shown in figure. The icosahedron I2 has a boron-carbon mixed-occupancy site B,C6 whose occupancy is B/C=0.58/0.42. Remaining 3 boron-carbon mixed-occupancy sites are bridge sites; C and Si sites are also bridge sites.

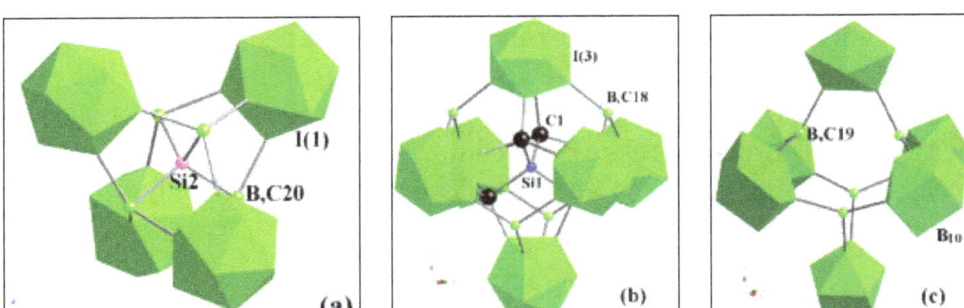

(a) Supertetrahedron T(1), (b) superoctahedron O(1) and (c) octahedral arrangement of the B_{10} polyhedra in the $Sc_{0.83-x}B_{10.0-y}C_{0.17+y}Si_{0.083-z}$ crystal structure.

More than 1000 atoms are available in the unit cell, which is built up by large structure units such as two supertetrahedra T(1) and T(2) and one superoctahedron O(1). As shown in figure, T(1) consists of 4 icosahedra I(1) which have no direct bonding but are bridged by four B and C20 atoms. These atoms also form tetrahedron centered by the Si2 sites. The supertetrahedron T(2) that consists of 4 icosahedra I(2) is the same as shown in figure (b); its mixed-occupancy sites B and C6 directly bond with each other. The superoctahedron O(1) consists of 6 icosahedra I(3) and bridge sites B, C18, C1 and Si1; here Si1 and C1 exhibit a tetrahedral arrangement at the center of O(1). The B_{10} polyhedra also arrange octahedrally, without the central atom, as shown in figure where the B and C19 atoms bridge the B_{10} polyhedra to form the octahedral supercluster of the B_{10} polyhedra.

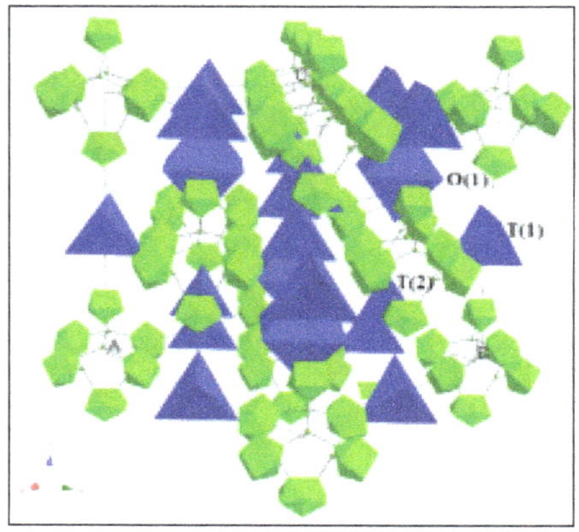

Boron framework structure of $Sc_{0.83-x}B_{10.0-y}C_{0.17+y}Si_{0.083-z}$ depicted by supertetrahedra T(1) and T(2), superoctahedron O(1) and the superoctahedron based on B_{10} polyhedron. Vertexes of each superpolyhedron are adjusted to the center of the constituent icosahedra, thus the real volumes of these superpolyhedra are larger than appear in the picture.

Using these large polyhedra, the crystal structure of $Sc_{0.83-x}B_{10.0-y}C_{0.17+y}Si_{0.083-z}$ can be described as shown in figure. Owing to the crystal symmetry, the tetrahedral coordination between these superstructure units is again a key factor. The supertetrahedron T(1) lies at the body center and at the edge center of the unit cell. The superoctahedra O(1) locate at the body center (0.25, 0.25, 0.25) of the quarter of the unit cell. They coordinate tetrahedrally around T(1) forming a giant tetrahedron. The supertetrahedra T(2) are located at the symmetry-related positions (0.25, 0.25, 0.75); they also form a giant tetrahedron surrounding T(1). Edges of both giant tetrahedra orthogonally cross each other at their centers; at those edge centers, each B_{10} polyhedron bridges all the super-structure clusters T(1), T(2) and O(1). The superoctahedron built of B_{10} polyhedra is located at each cubic face center.

Scandium atoms reside in the voids of the boron framework. Four Sc1 atoms form a tetrahedral arrangement inside the B_{10} polyhedron-based superoctahedron. Sc2 atoms sit between the B_{10} polyhedron-based superoctahedron and the O(1) superoctahedron. Three Sc3 atoms form a triangle and are surrounded by three B_{10} polyhedra, a supertetrahedron T(1) and a superoctahedron O(1).

$ScB_{14-x}Cx$ (x = 1.1) and $ScB_{15}C_{1.6}$

$ScB_{14-x}C_x$ has an orthorhombic crystal structure with space group *Imma* (No. 74) and lattice constants of $a = 0.56829(2)$, $b = 0.80375(3)$ and $c = 1.00488(4)$ nm. The crystal structure of $ScB_{14-x}C_x$ is isotypic to that of $MgAlB_{14}$ where Sc occupies the Mg site, the Al site is empty and the boron bridge site is a B/C mixed-occupancy site with the occupancy of B/C = 0.45/0.55. The occupancy of the Sc site in flux-grown single crystals is 0.964(4), i.e. almost 1. Solid-state powder-reaction growth resulted in lower Sc site occupancy and in the resulting chemical composition $ScB_{15}C_{1.6}$. The B-C bonding distance 0.1796(3) nm between the B/C bridge sites is rather long as compared with that (0.15–0.16 nm) of an ordinary B-C covalent bond, that suggests weak bonding between the B/C bridge sites.

$Sc_{4.5-x}B_{57-y+z}C_{3.5-z}$

$Sc_{4.5-x}B_{57-y+z}C_{3.5-z}$ (x = 0.27, y = 1.1, z = 0.2) has an orthorhombic crystal structure with space group *Pbam* (No. 55) and lattice constants of a = 1.73040(6), b = 1.60738(6) and c = 1.44829(6) nm. This phase is indicated as $ScB_{12.5}C_{0.8}$ (phase IV) in the phase diagram of figure. This rare orthorhombic structure has 78 atomic positions in the unit cell: seven partially occupied Sc sites, four C sites, 66 B sites including three partially occupied sites and one B/C mixed-occupancy site. Atomic coordinates, site occupancies and isotropic displacement factors.

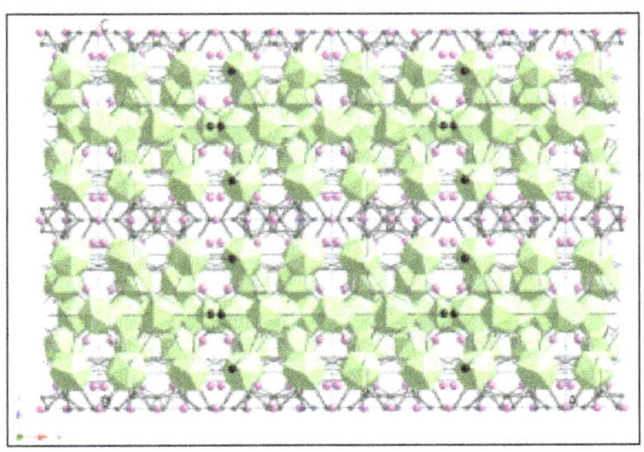

$Sc_{4.5-x}B_{57-y+z}C_{3.5-z}$ crystal structure viewed along the direction; 2.5 and 2 unit-cell sizes are depicted along the x-axis and z-axis, respectively.

More than 500 atoms are available in the unit cell. In the crystal structure, there are six structurally independent icosahedra I1–I6, which are constructed from B1–B12, B13–B24, B25–B32, B33–B40, B41–B44 and B45–B56 sites, respectively; B57–B62 sites form a B_8 polyhedron. The $Sc_{4.5-x}B_{57-y+z}C_{3.5-z}$ crystal structure is layered, as shown in figure. This structure has been described in terms of two kinds of boron icosahedron layers, L1 and L2. L1 consists of the icosahedra I3, I4 and I5 and the C65 “dimer”, and L2 consists of the icosahedra I2 and I6. I1 is sandwiched by L1 and L2 and the B_8 polyhedron is sandwiched by L2.

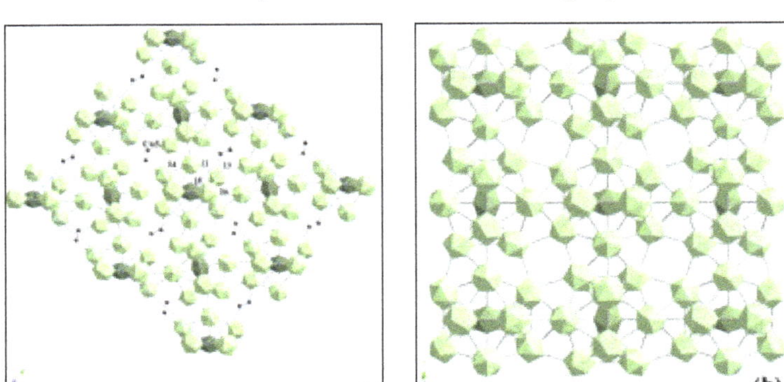

Two-dimensional presentation of supericosahedron connection in (a) $Sc_{4.5-x}B_{57-y+z}C_{3.5-z}$ and (b) YB_{66}. The central icosahedra of the supericosahedron are dark green.

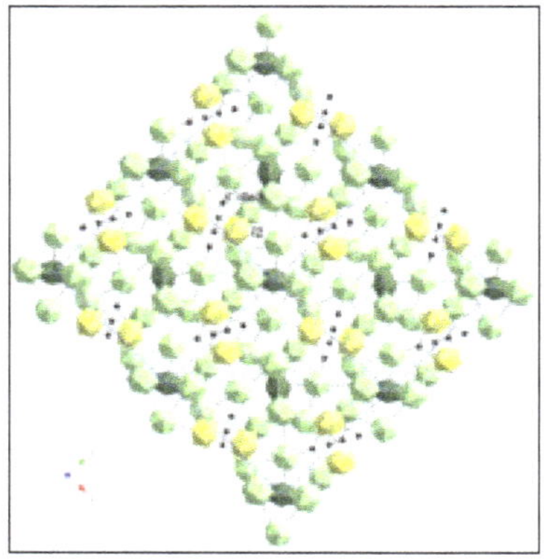

Locations of the I2 icosahedron (yellow) in the 2-dimensional supericosahedron framework of $Sc_{4.5-x}B_{57-y+z}C_{3.5-z}$.

An alternative description is based on the same $B_{12}(B_{12})_{12}$supericosahedron as in the YB_{66} structure. In the YB_{66} crystal structure, the supericosahedra form 3-dimensional boron framework as shown in figure. In this framework, the neighboring supericosahedra are rotated 90° with respect to each other. On the contrary, in $Sc_{4.5-x}B_{57-y+z}C_{3.5-z}$ the supericosahedra form a 2-dimensional network where the 90° rotation relation is broken because of the orthorhombic symmetry. The planar projections of the supericosahedron connection in $Sc_{4.5-x}B_{57-y+z}C_{3.5-z}$ and YB_{66} are shown in figures, respectively. In the YB_{66} crystal structure, the neighboring 2-dimensional supericosahedron connections are out-of-phase for the rotational relation of the supericosahedron. This allows 3-dimensional stacking of the 2-dimensional supericosahedron connection while maintaining the cubic symmetry.

The B_{80} boron cluster occupies the large space between four supericosahedra as described in the REB_{66} section. On the other hand, the 2-dimensional supericosahedron networks in the $Sc_{4.5-x}B_{57-y+z}C_{3.5-z}$ crystal structure stack in-phase along the *z*-axis. Instead of the B_{80} cluster, a pair of the I2 icosahedra fills the open space staying within the supericosahedron network, as shown in figure where the icosahedron I2 is colored in yellow.

All Sc atoms except for Sc3 reside in large spaces between the supericosahedron networks, and the Sc3 atom occupies a void in the network as shown in figure. Because of the small size of Sc atom, the occupancies of the Sc1–Sc5 sites exceed 95%, and those of Sc6 and Sc7 sites are approximately 90% and 61%, respectively.

$Sc_{3.67-x}B_{41.4-y-z}C_{0.67+z}Si_{0.33-w}$

$Sc_{3.67-x}B_{41.4-y-z}C_{0.67+z}Si_{0.33-w}$ (x = 0.52, y = 1.42, z = 1.17 and w = 0.02) has a hexagonal crystal structure with space group *P6m2* (No. 187) and lattice constants $a = b = 1.43055(8)$ and $c = 2.37477(13)$ nm. Single crystals of this compound were obtained as an intergrowth phase in a float-zoned single crystal of $Sc_{0.83-x}B_{10.0-y}C_{0.17+y}Si_{0.083-z}$. This phase is not described in the phase diagram of figure because it is a quaternary compound. Its hexagonal structure is rare and has 79 atomic positions in the unit cell: eight partially occupied Sc sites, 62 B sites, two C sites, two Si sites and six B/C sites. Six B sites and one of the two Si sites have partial occupancies. The associated atomic coordinates, site occupancies and isotropic displacement factors.

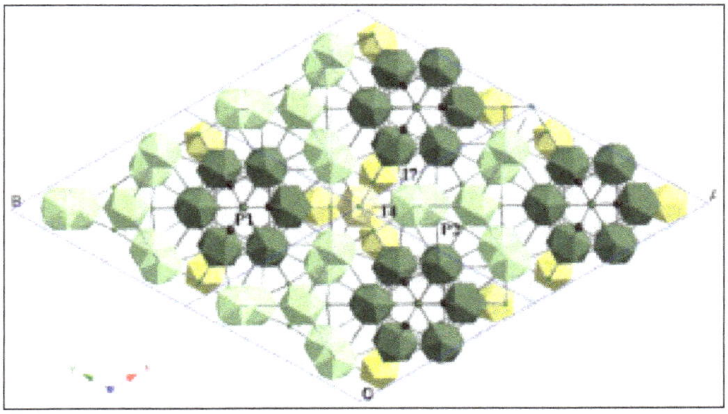

Boron framework structure of $Sc_{3.67-x}B_{41.4-y-z}C_{0.67+z}Si_{0.33-w}$ viewed along the *c*-axis.

There are seven structurally independent icosahedra I1–I7 which are formed by B1–B8, B9–B12, B13–B20, B/C21–B24, B/C25–B29, B30–B37 and B/C38–B42 sites, respectively; B43–B46 sites form the B_9 polyhedron and B47–B53 sites construct the B_{10} polyhedron. B54–B59 sites form the irregularly shaped B_{16} polyhedron in which only 10.7 boron atoms are available because most of sites are too close to each other to be occupied simultaneously. Ten bridging sites C60–B69 interconnect polyhedron units or other bridging sites to form a 3D boron framework structure. One description of the crystal structure uses three pillar-like units that extend along the *c*-axis that however results in undesired overlaps between those three pillar-like units. An alternative is to define two pillar-like structure units. Figure shows the boron framework structure of $Sc_{3.67-x}B_{41.4-y-z}C_{0.67+z}Si_{0.33-w}$ viewed along the *c*-axis, where the pillar-like units P1 and P2 are colored in dark green and light green respectively and are bridged by yellow icosahedra I4 and I7.

These pillar-like units P1 and P2 are shown in figures, respectively. P1 consists of icosahedra I1 and I3, an irregularly shaped B_{16} polyhedron and other bridge site atoms where two supericosahedra can be seen above and below the B_{16} polyhedron. Each supericosahedron is formed by three icosahedra I1 and three icosahedra I3 and is the same as the supericosahedron O(1) shown in figure. The P2 unit consists of icosahedra I2, I5 and I6, B_{10} polyhedron and other bridge site atoms. Eight Sc sites with occupancies between 0.49 (Sc8) and 0.98 (Sc1) spread over the boron framework.

As described above, this hexagonal phase originates from a cubic phase, and thus one may expect a similar structural element in these phases. There is an obvious relation between the hexagonal *ab*-plane and the cubic (111) plane. Figures show the hexagonal (001) and the cubic (111) planes, respectively. Both network structures are almost the same that allows intergrowth of the hexagonal phase in the cubic phase.

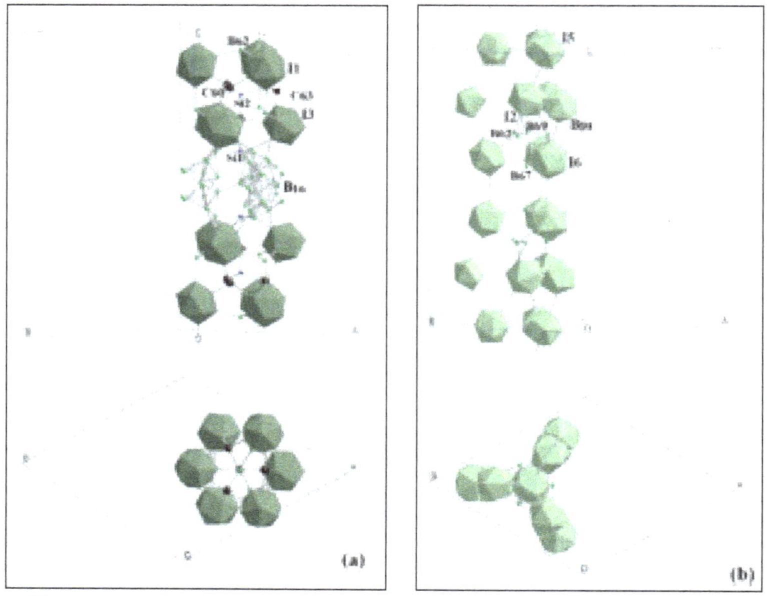

(a) Pillar-like structure unit P1 that consists of icosahedra I1 and I3, irregularly shaped B_{16} polyhedron and other bridge site atoms. (b) Pillar-like structure unit P2 that consists of icosahedra I2, I5 and I6, B_{10} polyhedron and other bridge site atoms.

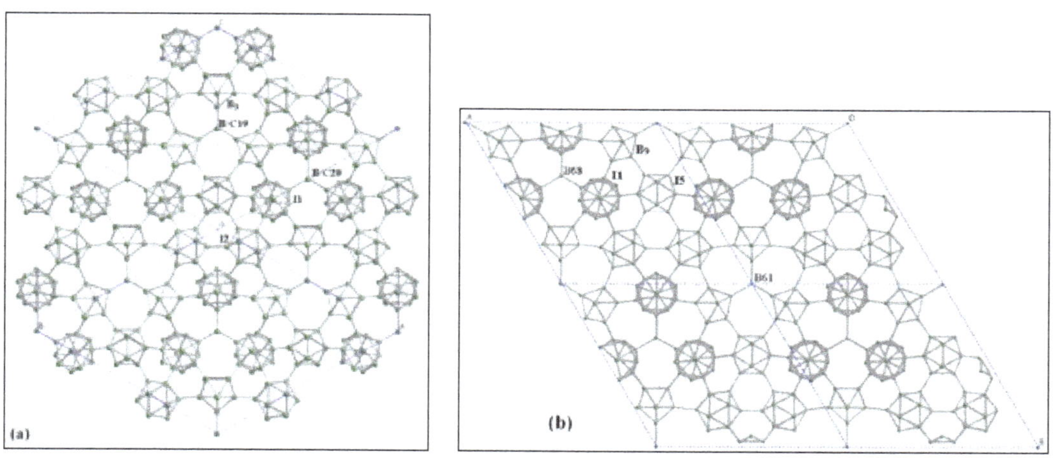

(a) The sliced (111) network structure of the cubic phase, and (b) the (001)-oriented boron network layer of the hexagonal phase.

Applications

The diversity of the crystal structures of rare-earth borides results in unusual physical properties and potential applications in thermopower generation. Thermal conductivity of boron icosahedra based compounds is low because of their complex crystal structure; this property is favored for thermoelectric materials. On the other hand, these compounds exhibit very low (variable range hopping type) p-type electrical conductivity. Increasing the conductivity is a key issue for thermoelectric applications of these borides.

YB_{66} is used as a soft-X-ray monochromator for dispersing 1–2 keV synchrotron radiation at some synchrotron radiation facilities. Contrary to thermoelectric applications, high thermal conductivity is desirable for synchrotron radiation monochromators. YB_{66} exhibits low, amorphous-like thermal conductivity. However, transition metal doping increases the thermal conductivity twice in $YNb_{0.3}B_{62}$ as compared to undoped YB_{66}.

Evolutionary Algorithms in Crystal Structure Analysis

Evolutionary algorithms are "probabilistic" optimization procedures. Differently from the "deterministic" ones (yielding an unique and reproducibile solution for assigned input data), EA are somewhat aleatory and not reproducible. Such methods are part of the heuristics. To be pedantic the term algorithm should indicate some sequence of mathematical operations producing a foreseeable result if applied to definite input data. There is a resounding oxymoron in the terms evolutionary algorithm. This contradictory language is however diffuse; as this one is not an erudite essay of epistemology, we prefer to be tolerant and fly over on semantic questions. Really evolutionary methods have aleatory routes and also aleatory conclusions in many cases (e.g. stock exchange estimate, weather forecast, etc.); in some cases instead (it happens in crystallography), unique and certain results can be reached in spite of the multiplicity of routes travelled.

Multi-parameter optimization problems are encountered in many fields: industrial process planning, financial investments, environment control, hurricane evolution, and many others. We shall however not consider the general aspects of EA but rather to treat about a specific argument: the study of chemical structures based on X-ray diffraction. In this field, as shown later, there is a lot of a priori and useful information so that the aleatority of the procedure can be considerably attenuated. Evolutionary methods in crystallography are better formulated as Constraint Satisfaction Problems.

Evolutionary methods have a future in crystallography, particularly in solving hard problems. The structural elucidation from X-ray powder diffraction data is among these. However, procedures are not consolidated yet, there are numerous alternatives to be considered for which the common sense and the experience, rather than the theory, could have a role.

The effectiveness of EA in solving crystal structures by diffraction methods is improved if one makes systematically use of the so called internal coordinates (i.c.) for describing structures, instead of the more common cartesian coordinates. The i.c. were first considered, in diffraction analysis. A computer program performing least-square refinement of i.c. (LALS, Linked Atom Least Square) was first issued and later updated repeatedly. LALS has been used mainly in studying polymers. Also Tadokoro discussed the use of i.c. in polymer case.

Both Arnott (and disciples) and Tadokoro did however neglect a very important complication: the mutual independency (non-redundancy) of i.c., being of great importance in all crystallogaphic applications, and in particular in EA, as will be shown later.

Now a procedure for carrying-out a molecular building always using non-redundant i.c. has been devised; it is very simple, practical, and perfectly analytical, runs thoroughly, and eliminates the necessity of using exotic instruments in matrix manipulations (pseudoinversion, diagonalization, etc.; note that the mathematic instruments employed are those of Newton, Gauss, and Lagrange).

Crystalline Matter and X-ray Diffraction

Characterization of chemicals, both natural and synthetic, is done always looking at the matter at the atomic scale. In 19th century, atoms were speculative entities and scientists had not instruments for ascertaining their shape or position. Nevertheless, in the middle of century, some of them, with evident attitude versus the heuristics (without awareness of course as heuristics did not exist yet) were persuaded about molecules (proposed many years before by Avogadro finding much skepticism); they begun, with much imagination and using logical arguments, to give them definite shapes, for instance the tetrahedral shape to methane and the hexagonal shape to benzene. Many of these intuitions were demostrated later be perfectly exact. Today to ascertain the geometrical structure of molecules is not more a work of imagination: it is an exact science. Structure is a "property" of any pure substance, defined and reproducible, like colour, density, melting temperature, etc.

The discovery of X-ray diffraction by crystals by Laue in 1912 made possible to ascertain the structure. In the crystalline state, necessarily solid, atoms are ordered: there is a small portion of space, a parallelepiped, generally oblique, termed the unit cell, which repeats itself by translation in three directions and generates the whole solid. If the disposition of atoms in the cell is determined, the whole crystal is determined. Frequently there is also symmetry: only a fraction of the unit cell is independent; the remainder if given by appropriate orthogonal transforms. Edges and interaxial angles of the unit cell are collectively termed lattice constants. Just order makes possible X-ray diffraction of crystal, whose study permits to guess the matter at the atomic scale; resolution is of the order of 0.02 Å or better; consider that the separation of bonded atoms is 1-2 Å. The size of molecules studied by X-ray diffraction was initially modest; now studying 50-100 atom molecules is a common matter.

Molecules and Molecular Crystals

Molecules are finite assemblies of atoms joined by strong forces (chemical bonds). In molecular crystals molecules are orderly assembled through well weaker non-bonding forces (van der Waals forces, dipole-dipole forces, hydrogen bonds, etc.).

Crystals diffract X-ray radiation: when a beam of monochromatic radiation strikes a crystalline sample many beams emerge and their intensities can be one-by-one measured. The intensity of the diffracted beams is related to the molecular structure.

In absence of symmetry, if V is the volume of the unit cell, λ the radiation wavelength, and $2\vartheta m$ the highest angular deviation, there are altogether $3.35(V/\lambda^3)\sin 3\vartheta m$ different diffracted beams (reflections). The above relationship indicates that the number of reflections raises as $\sin \vartheta m/\lambda$ increases. The latter is an important parameter: the higher is $\sin \vartheta m/\lambda$ the higher is the chance of success in a structural analysis, and the higher is the accuracy of the result. Consider however that the intensity of the reflections decreases as ϑ angle increases and that $\sin \vartheta m/\lambda$ may be limited by natural circumstances: there are materials which are intrinsecally weakly diffracting; their structural study cannot be done accurately.

Structure and Diffraction

It is possible to obtain crystal structures from X-ray diffraction data thanks to the mathematical relationship between atomic positions and the intensities of the reflections. The latter, as numerous as the volume of the unit cell is large, are singled out by three integer numbers: the Bragg indices (h, k, ℓ). Each is deviated from the incident direction by a characteristic angle 2ϑ, an analytical function of the Bragg indices and of the lattice constants. One must intercept these beams and measure intensity.

There are nowadays very sophisticated instruments (diffractometers), controlled by computers, allowing, with a minimal human intervention and in a short time, a complete characterization of a crystal: measurement of the lattice constants, diagnosis of symmetry, localization and measurement of all reflections with their Bragg indices h, k, ℓ. If the unit cell contains N atoms, and x_j, y_j, z_j are the atomic coordinates (referred to the unit cell), the complex quantities:

$$F(h,k,\ell)=\sum_{j=1}^{N} f_j \exp\left[2\pi i\left(h\frac{x_j}{a}+k\frac{y_j}{b}+\ell\frac{Z_j}{c}\right)\right]$$

are termed structure factors. The f_j (atomic factors) are known real quantities which depend on the chemical nature of the j-th atom and on $\sin \vartheta/\lambda$. The larger is the atomic number the higher are f_j. In every cases f_j decreases as $\sin \vartheta/\lambda$ increases.

The above relationship applies at 0 K; at higher temperature F's are somewhat reduced because of the thermal vibration. There are of course as many $F(h, k, \ell)$ as reflections.

Now the squared moduli of $F(h, k, \ell)$ should be orderly proportional to the measured diffraction intensities $I(h, k, \ell)$. This occurs, of course, when the x_j, y_j, z_j coordinates are the true ones. To get the unknown crystal structure one must find all the x_j, y_j, z_j rendering the one can assume that the solution is the one rendering χ^2 a minimum. For this reason to find a structure is to solve a global optimization problem. The above χ^2, divided by the sum of $F^2(h, k, \ell)$ is the common R^2 index, widely used as fitness function.

Single-crystals and Polycrystalline Samples

X-ray diffraction studies use either single-crystal samples or polycrystalline samples (powder, fiber). In the second case things are more difficult because all diffracted beams having a given 2ϑ are overlapped. While the diffraction pattern of a single crystal is a three-dimensional function $I(h, k, \ell)$ rich of information, the powder diffraction is an uni-dimensional function $I(2\vartheta)$ with poor information.

Structural studies based on powder diffraction data meet the difficulty of measuring the diffracted intensities one-by-one, singled out by the Bragg indices, in presence of overlap. The difficulty can be overcome employing deconvolution techniques or, alternatively, renouncing to the separation of reflections and considering the quasi-continous function $I(2\vartheta)$ exploiting the so called full-pattern powder profile analysis.

In this case, the measured diffraction intensities $I(2\vartheta_i)$ are compared with the computed ones given by:

$$I_{calcd,i} = \sum_k F_k^2 \Omega(2\vartheta_i - 2\vartheta_k)$$

where the sum is extended to all the reflections whose Bragg $2\vartheta_k$ fall near to the current $2\vartheta_i$ point. In equation $I_{calcd,i} = \sum_k F_k^2 \Omega(2\vartheta_i - 2\vartheta_k)$ $\Omega(2\vartheta_i - 2\vartheta_k)$ is the peak function having a maximum for $2\vartheta_i = 2\vartheta_k$. In full-pattern profile analysis the residual χ^2 considered is given by the expression:

$$\chi^2 = \sum_{2\vartheta_i} \left[I(2\vartheta_i) - I_{calcd,i} \right]^2$$

and the most used fitness index is $Rwp = \chi^2 / \sum I^2(2\vartheta)$.

Just because EA techniques need a reduced number of data they are attractive in studies based on powder diffraction, particularly when the full-pattern profile analysis is performed.

When a structure has been solved (at a coarse level) a second problem arises: to refine atomic positions. This is a local optimization problem. Once again the goal is to find the minimum of χ^2 varying systematically atomic positions in the vicinity of the initial coarse values and adding some parameters for expressing the thermal vibration.

While the local optimization is done using deterministic algorithms (typically a least-square refinement) and proceeds, in most cases, without problems, the global optimizazion, even at coarse level, is much more exacting.

The remainder of this chapter concerns only global optimization problems. In addition we shall not treat the traditional solution methods but only some based on evolutionary algorithms.

The Three-dimensional Space and the Problem space

Solving a structure of N atoms means finding N positions in the 3D space. Unfortunately there are no algorithms for finding positions one by one (unless rather simple cases occurring in inorganic chemistry). Atomic positions must be find simultaneously.

An alternative way to define the terms of problem is to refer to another space, the so called problem space, having 3N dimensions instead of three. To catch a structure means to locate a single point in this hyper-space. Simple to say, but very difficult to do. As explained later the EA follow this crazy idea.

Small-medium size problems, are solved using deterministic algorithms, provided that the available diffraction measurements are numerous, accurate, and complete (i.e. including all the intensities with the diffraction angles ϑ lower than the assigned $2\vartheta_m$).

One could ingenuously believe that all structural problems can be solved moving the representative point systematically in the whole problem-space and examining all solutions. Atomic coordinates are however real and continuous quantities; so points are infinite. Well, even though finite (and coarse) intervals are considered to render the problem-space discontinuous, one obtains astronomic numbers also in the simple cases. The systematic exploration of the whole problem-space is not a practical tool, unless some tricks are adopted. Possible strategies are:

- To reduce substantially the number of variables adopting other kinds of coordinates rather than crystallographic and choosing coordinate systems for which a good fraction of variables are predictable.
- To render the problem-space discontinuous.
- To reduce as much as possible the range of variation for each variable.

We shall suppose that the crystal under study is a molecular crystal with known composition (chemical formula) and known interatomic connection (structural formula); consider that the chemical formula is obtained by chemical analysis; structural formula is the result of a number of physical observations (e.g. NMR, IR, UV spectroscopy) and chemical observations (relations of the unknown substance to other already known).

Using diffraction techniques one can establish size, shape, and symmetry of the unit cell. From the experimental density of crystals one obtains the atom content of the unit cell. Using the diffractometers one obtains an appropriate number of diffracted intensities.

Coordinates

Structures are commonly described by using the crystallographic coordinates (c.c.) to define atom positions. The c.c. (collectively indicated with pi) are 3N if atoms are N.

As a matter of fact, infinite alternative systems for describing structures are conceivable: if new coordinates qi are introduced, so that one can transform from pi to qi and viceversa using analytic and biunivocal relationships, the qi can be used as alternative stuctural variables. The number of qi must be, of course, again 3N.

Which are the alternative useful descriptions among the infinite possible ones? To reply it is important to point out that atomic positions are not fully impredictable. The systematic study of molecular crystals did demonstrate that atomic positions are not random quantities. Any structural hypothesis must obey to the following rules:

- Atoms must be appropriately separated among them; for each atom pair distance is anything but random: if the two atoms are chemically bound, atom separation is a bond-length (b-l) and must be close to the sum of the atomic radii; if not bound and spaced by three bonds or more, the distance must be higher than the sum of the so called van der Waals radii; the same applies when the two atoms belong to different molecules (packing distances).
- The Bond-Angles (b-a) must be close to the "canonical" values prescribed by the rules of orbital hybridation: angles close to 109.5 ◦ on carbon atoms with sp3 hybridation, close to 120 ◦ on carbon atoms with sp 2 hybridation, etc. The examination of "molecular models" suggest in much cases angles in rather restricted intervals.
- The molecular conformation must obey to the rules of stereochemistry: aromatic rings must have D6d symmetry, the sequences C–C=C–C have to be planar, the torsion-angles (t-a) about single bonds must have values close to those typical of ethane-like molecules (-60, 60, 180 ◦), etc. These limitations can be thus "rigid" in some cases (double bonds), "flexible" in other cases (single bonds).

The above restrictions for b-l, b-a, and t-a are of simple mathematical formulation if the crystallographic coordinates pi are employed; there is however the disadvantage that the restrictions (constraints) apply not to the pi themselves, but to a number of mathematical functions of the pi . Indeed there are procedures for doing minimizations in presence of constraints: the Lagrange method with the drawback that only a few constraints can be accounted for, while in molecular building there are numerous. Instead, introducing alternative coordinates qi chosen with cleverness, it is possible to impose the above constraints not on the functions but directly on the qi . If so happens one simply removes the qi from the list of variables.

The Internal Coordinates: Eyring Algorithm

It is customary studying complicate problem by means of the so called internal coordinates (i.c.), used in performing molecular building in various other contexts (e.g. spectroscopy, theoretical chemistry). The pioneer work (in chemistry) was that of Eyring who devised the procedure later called Z-matrix . Really the idea is well older; it was skilfully described by Lagrange who coined the term generalized coordinates. Lagrange studied not molecules but machines; things are rather similar after all: Lagrange's machines are made by rigid objects connected to each other with limited freedom of movement; in molecules the rigid objects are the chemical bonds and movements are limited to torsion and (in small amount) to bending. Eyring's algorithm is simple, provided that there are neither rings nor polyhedra. The idea is to build molecules first defining three atoms, then inserting the other atoms stepwise as function of three i.c. in each step: a b-l b, a b-a τ and a t-a ϑ. Of course, in each step one must specify which atoms are to be considered. Three i.c. are needed for starting a molecular building, viz. two b-l (b1 , b2) and a b-a (τ3). It is customary to put the 1st atom on the origin, the 2nd atom along x axis, the 3rd atom in the x, y plane with y > 0.

Figure shows how the building starts (chbe command, green atoms) and the Eyring construction (red atom). The construction of the 5th atom (blue) is discussed later.

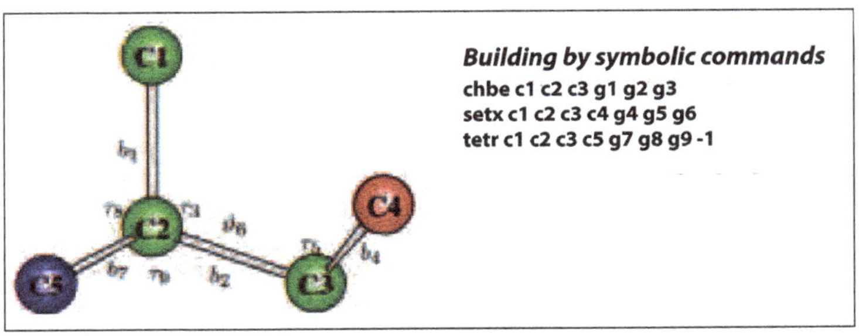

Building a 5 atoms molecule.

The Eyring's machinery for attaching a new atom after a given sequence A, B, C, is as follows: one computes the vector v with components b cos τ, b sin τ cos ϑ, b sin τ sin ϑ, the vector product u =(B – A) × (C – B), and the orthogonal transformation matrix T aligning A and B points along z ′ and making y ′ C = x ′ C = 0; then adds to C atom the product T · v . The analytical expression of T matrix is:

$$T = \begin{vmatrix} x_B - x_A & x_C - x_B & u_x \\ y_B - y_A & y_C - y_B & u_y \\ z_b - Z_A & Z_C - z_B & u_z \end{vmatrix}$$

with each column normalized to length 1.

Besides the Eyring's construction (set x) needing three variables b, τ, ϑ, another construction is practical when atoms are inserted on tertiary carbon atoms of known chirality; this construction (tetr) makes use of two b-a τ_1, τ_2 (instead of one b-a and one t-a); the advantage is the restricted range of the b-a compared with the wide range of the t-a. The above 5-atom skeleton has been built just with 3 constructions: chbe (for atoms C1–C2–C3, in green), setx (for atom C4, in red), and tetr (for atom C5, in blue).

In figure the three constructions are indicated "symbolically" according to a conventional syntax.

Of course six other variables must be added to the molecular i.c. for defining the actual position and orientation of the molecule in the unit cell: three translations and three rotation angles. Rototranslation i.c. can be less than six because of crystal symmetry.

Redundant Coordinates

A serious drawback encountered in building molecules using Eyring method is the possible redundancy of the i.c. In simpler words they might be not independent. While independent i.c. can be varied each in turn preserving the structural formula, this does not apply in case of redundancy. When molecular building is performed within random search and genetic procedures, the preservation of the structural formula is of basic importance. Also regarding reduncancy there is nothing new after Lagrange; he said all the necessary: the generalized coordinates are exactly the same thing.

As illustrated below, non-redundancy in molecular building can be fulfilled by selecting the 3N −6 i.c. of a molecule with shrewdness among b-l, b-a, t-a, and also bending angles if necessary. Bending angles become necessary in two cases: cyclic molecules and polyhedric molecules.

Cyclic Molecules

A possible solution overcoming redundancy in cyclic molecules is shown in considering the case of cyclohexane. Since the g_i must be, at a molecular level, 3N – 6, considering N bond-lengths, only other 2N – 6 g_i must be assigned, necessarily, angular. The method adopted integrates Eyring's procedure for building atoms C1, C2, C5(9 g_i are employed altogether, 4 b-l, 3 b-a, and 2 t-a) with a new machinery for building atom C6 using two b-l and one angle only, a so called bending angle (ϕ); with this construction the molecule is bent about a line crossing two atoms separated by one bond only. Such construction (termed flap) has been added to the above ones. The machinery for computing C6 as a function of ϕ and the two b-l C1–C6 (g11) and C5–C6 (g12) is very simple. The construction shown in figure, termed flap resolves the problem for rings of any dimension and also every polyciclic molecule (e.g. decaline, steroids, etc.) and resolves also intricate multicyclic molecules like norbornane, pinene, spirocompounds, etc.

The construction shown in figure, termed flap resolves the problem for rings of any dimension and also every polyciclic molecule (e.g. decaline, steroids, etc.) and resolves also intricate multicyclic molecules like norbornane, pinene, spirocompounds, etc.

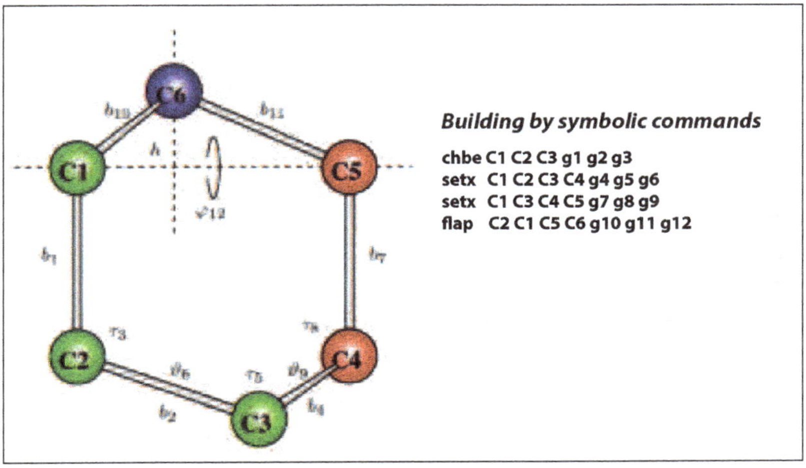

Building a ring without redundancy, cyclohexane C_6H_6,
(the blue atom is computed by a flap instruction).

Polyhedric Molecules

A possible solution overcoming redundancy in polyhedric molecules is shown in figure considering the simplest polyhedron: the cubane C_8H_8, represented as a bare carbon atom skeleton. Remember that the g_i must be, at the molecular level, $3 \times 8 - 6 = 18$ and 12 g_i are b-l. In order to build without redundancy one must use only 6 angles. Building the first 7 atoms can be done as indicated symbolically in the figure. After the 5th command one atom is lacking, C8, and all the 6 angles have been used besides 9 over 12 bond-lengths. To complete the molecule one can find x_8, y_8, z_8 imposing C8 has assigned distances b16, b17, b18 from points C4, C5, C7. This is a simple (and classical) problem of 2nd degree having an unique solution provided that the polyhedron is convex. The command cage computes the lacking atom. Note that cage allows modeling also giant polyhedra like the fullerenes.

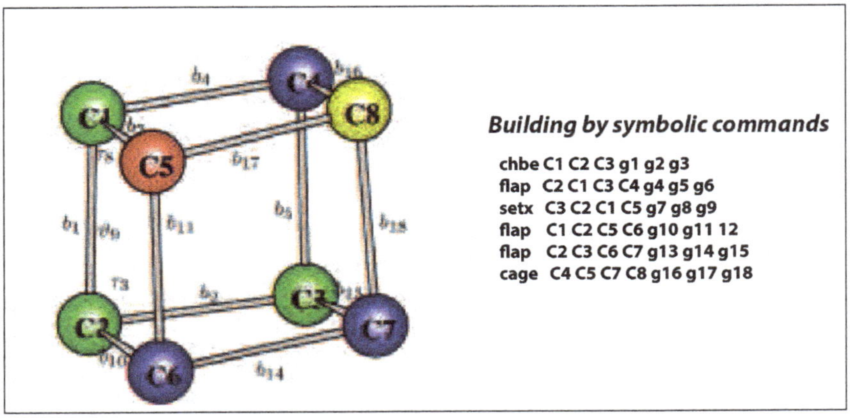

Building a polyhedric molecule without redundancy (cubane C_8H_8).
The green atoms are computed by chbe, the red atom by setx, the
blue atoms are computed by flap, and the yellow atom by cage.

Other Constructions and Symbolic Building

Molecular and crystal building through internal coordinates is not used only in performing structural analysis by EA. It can be used instead in all conventional procedures. In particular for doing interactive modeling, trial-and-error calculation, and at last for doing the least-square refinement of structures based on the i.c. themselves as optimized variables. In the latter case the non-redundancy of the variables plays a crucial role since only in absence of redundancy normal matrices are always non-singular.

For these reasons all the mentioned constructions have been programmed devising an unique subroutine (termed LAGR , an homage to Lagrange, the authentic, and ignored, discoverer of the i.c.) performing the whole construction at each call. To simplify the input data preparation LAGR has been structured so that the various constructions invoked are specified symbolically according to a conventional syntax with one line of data for each construction step. Examples of this syntax are indicated in the figures and in the input data for procedure validation.

Besides the mentioned four basic constructions needed for building molecules of any kind (chbe, setx, tetr, flap, and cage, tetr is really not indispensable) there are numerous other constructions, not strictly necessary but very useful. Among them there are phen (building phenyl groups), metl (building a CH_3 group) and others. LAGR subroutine includes also the commands necessary for placing the built molecule in the unit cell, for doing orthogonal transforms, for referring the molecule to its inertial axes, and others. A clever usage of these commands brings to a further reduction of the effective number of i.c. by imposition of local symmetry, i.e. the identity of structurally independent i.c. but chemically indistinguishable. At last a number of special commands were implemented for the case of linear polymers: to create regular helices and chains with glide-planes, to orient the chain parallel to the crystallographic c edge, etc.

General Layout of a EA based Procedure: Metaheuristic Approach

The evolutionary (or genetic) algorithms for resolving global optimization problems are based on the idea of finding the "true" solution starting from a number of more or less arbitrary solutions and performing combinations of the variables analogous to the ones occurring in the cellular reproduction.

There is an analogy between chromosomes and molecular structures. As a chromosome can be described by a sequence of genes, a molecular structure can be described by a sequence of structural variables. For these reasons the terms are used as synonyms.

In order to limit the aleatority of the procedure it is practical to follow the so called metaheuristic approach assuming that the g_i vary as multiple of a small increments. With this contrivance the i.c. (real and continuous quantities) become integer numbers whose upper limit is limited by the range of excursion of the g_i themselves. "Solutions" can binary-coded using a finite number of bits. Consider anyway that the number of points in the problem-space remains astronomic; for instance a 10-variable structure codified in 60-bits may assume $2^{60} \approx 10^{18}$ values. If, for simplicity, a 5-gene case is considered, and 7, 6, 3, 4, 6 bits are dedicated to the 5 genes respectively, the binary code will be (braces groups the bits of given g):

$$\underbrace{b_6b_5b_4b_3b_2b_1b_0}_{g1}\ \underbrace{b_5b_4b_3b_2b_1b_0}_{g2}\ \underbrace{b_2b_1b_0}_{g3}\ \underbrace{b_3b_2b_1b_0}_{g4}\ \underbrace{b_5b_4b_3b_2b_1b_0}_{g5}$$

Planning the Complete Structural Building

It is propedeutical to write-up a complete schema for the structural building based on the symbolic language. This important step is anything but "automatic"; rather it requires much attention and cleverness. One must assign the non-redundant i.c. and prepare the full list of the symbolic building commands which, using the aforementioned subroutine LAGR, will provide the full structural building as function of the current i.c. This schema serves not only for carrying out the EA procedure, but for all the numerous options of TRY program (calculation of structure factor, model adjustment, crystal packing, least-square refinement, geometrical computations, etc.). Among the symbolic instructions there is one indispensable for carrying out the random search in the problem space, termed xcon, and discussed in details later.

Binary Encoding

Now the user must: i) decide which g_i must be kept fixed and which varied (it is customary to exclude the bond-lengths and, possibly, also the bond-angles); ii) assign to each variable g_i a convenient increment Δg_i (the g_i will be varied by multiple of Δg_i), iii) assign to each variable g_i how many bit (m_i) are to be used in binary encoding. Δg_i and mi determine the range of excursion of the g_i (range it is exactly $\Delta g_i \times 2^{mi}$).

It is evident that some common sense must be exerted in assigning Δg_i and m_i with a compromise between fine steps in exploring the problem-space and wide spanning intervals. For instance, for a bond angle on a carbon atom one could assign $\Delta = 1.0^\circ$ and mi = 4; the angle will span a sufficient interval of 16°. Torsion angles (about single bonds), have instead higher uncertainity; giving them e.g. 5 bits and $\Delta g_i = 3^\circ$ will permit a span of 64°. A large number of bits, e.g. 7 or 8, must be assigned to the molecular rotation angles: using m = 7 and $\Delta = 2.8^\circ$ will ensure a 360° span; with m = 8 the same is fulfilled for finer step: 1.4°. Using common sense is no doubt necessary; but also experience turns out to be useful. In this mechanism the initial value of each variable $g^\circ i$ (to be defined in the input data) plays a role, since the values assumed by g_i are (k = 1, 2 . . . 2^m).

$$g_i^\circ\ ,\ g_i^\circ - \Delta_{gi},\ g_i^\circ + \Delta g_i^\circ, g_i^\circ\ -\ 2\Delta_{gi}, g_i^\circ + 2\Delta_i^\circ, g_i^\circ - 3\Delta_{gi}, \ldots$$

Note however that the g_i° are not critical quantities, provided they are internal to the span intervals.

Creating the Initial Population: Constraints

The first step of an evolutionary procedure is the creation of a population of tentative structures by means of a Montecarlo method (using e.g. pseudo-random numbers).

One must assign the size of the population desired (e.g. 100-300 items) and the selection criteria. Size and selection criteria control, in a very unforeseeable manner, the duration of the search, from few minutes to days. Anyway the search can be in any moment interrupted and restarted with new parameters, either creating a new list of solutions or queuing to the existing one.

If K is the number of g_i searched (K is of course well lower than 3N), the random search is performed generating, for each tentative structure, K integer random numbers each in the range $0 \div 2^{mj} - 1$ to obtain a random combination of gi . By using the cited subroutine LAGR, the corresponding model is created. It is evident that most of these models will be unfeasible. There are two alternatives: i) to retain all solutions, attaching to each solution a proper "penalty function", ii) to apply some selection criteria and reject immediately unfeasible items. The latter has been chosen in our procedure to avoid endless populations. Much experience is anyway necessary for calibrating the rejection parameters, actually under user control.

Selection Criteria: The xcon Command

The computer program implemented in our laboratory includes the following selection criteria (filters):

- The 1st filter, most incisive and mandatory, is the one checking the connectivity of the trial structure versus the known connectivity. We define as connectivity a group of eight integer codes (connv) to be inserted in the xcon command: the first is the number of atom-pairs separated by one bonds, the second the same number for two bonds, etc. up to 8 bonds.

- The 2nd filter eliminates structures with unfeasible bond angles. Non bonded pairs must have distances exceeding more than sepn2 (atoms separated by two bonds) or sepn3 (atoms separated by three or more bonds) the sum of covalent radii. Unrealistic structures with acute bond angles can so be eliminated. (In few cases, eg. cyclobutane, acute angles occur). The quantities sepn2 and sepn3 are supplied in the command xcon.

- The 3rd filter checks the connections of the trial molecule with the neighbour (in molecular crystals no connections should be present; in linear polymer, instead, two connections take place); the number of external connections is also assigned in xcon command. Connection are illegal if distance is lower than the sum of covalent radii augmented by tolnk, also supplied in xcon.

- The 4th filter consists in computing the lattice energy, again rejection occurs if energy exceeds the assigned threshold, chosen by the user at the beginning of search.
- The 5th filter is based on the R_2 index defined as the residual χ_2 divided by the sum of intensities. the tentative solution is rejected if R_2 exceeds an assigned value. Also for R_2 the threshold is chosen by the user.

Parameters sepn2, sepn3, tolnk, linkno and connv are defined in command xcon. Note that the same filters apply also in the subsequent breeding.

It is evident the importance of assigning the above values cum grano salis, avoiding both excess or lack of severity. Indeed the ratio accepted-structures/generated-structures leans to be low (e.g. 10^{-5}); the conformational freedom of the molecule has of course a critical role.

Optimizing of the Initial Population

We have introduced (as an optional) the local optimization of the random solutions found with the above procedure. Each random solution is a point in the (discontinuous) problem-space. The idea is to examine the nearest points in the problem-space and find points possibly more promising, based e.g. on R_2 index. In a K-dimension problem space points are $3^K - 1$, a value prohibitively high if K is large.

On belief that the local optimizazion is a good idea (experience seems to confirm) we have exploited a trick: to assign a reasonable value to the number of neighbouring points P to be examined, say 500-1000, and select, at random, P points among the $3^K - 1$ ones. With this contrivance the duration of an optimization step is reasonable.

Genetic Combinations

The heart of the genetic procedure is the systematic combination (breeding) of the solutions belonging to the initial population following the rules of genetics. In this context two kinds of combinations are considered: the cross-over and the mutation. A cross-over (binary combination) consists of an interchange of a selected gene belonging to the two mating items; a mutation (unary combination) is the change of a single bit (chosen at random) belonging to the bit-string. Considering a 9-gene example and a cross-over between genes g_2, g_7 one obtains:

$$\begin{aligned}
&\text{Parent sequence (father):} && g_1\ g_2\ g_3\ g_4\ g_5\ g_6\ g_7\ g_8\ g_9\\
&\text{Parent sequence (mother):} && g_1'\ g_2'\ g_3'\ g_4'\ g_5'\ g_6'\ g_7'\ g_8'\ g_9'\\
&\text{Child sequence:} && g_1\ g_7'\ g_3\ g_4\ g_5\ g_6\ g_2'\ g_8\ g_9\\
&\text{Mutated sequence:} && g_1\ g_7'\ g_3\ g_4^*\ g_5\ g_6\ g_2'\ g_8\ g_9
\end{aligned}$$

the symbol * has been used for a mutation (the g_4 has been chosen by chance). In each case a new sequence is born for each crossover, two if also a mutation takes place. If the crossover takes place for all the pairs of K genes, $K \times (K + 1)/2$ new strings occur; twice as many if also mutation is done.

Of course if two feasible structures are coupled, it is not warranted that child structures are meaningful. The above described "filtering" could be used again to eliminate unfeasible structures. At present the structures having R_2 higher than an assigned limit or a lattice energy higher than an assigned limit are eliminated. The limits, of course, should be fixed using experience. The new selected structures, anyway, are added to the initial ones.

The crossover can be done either considering a single gene, selected at random, or considering all genes; in the last case there are K combinations, for each pair of mated structures. When the genetic combination is ended the population of solution is extended by an amount depending on the used filters. The new population will be again sorted using an appropriate figure like the R_2 index.

Examining the sorted list one observes, frequently, that the first solutions happen to be very similar to each other. That is no doubt an indication of success. If this is not the case a good idea is to repeat the breeding as many times as it is necessary. If however the breeding has resulted ineffectual in finding new solutions, it is better to repeat the whole process in more appropriate conditions. The exploration of a wide fraction of the problem-space is no doubt essential for the success.

Procedure Validation

For testing purposes two already reported structures have been considered using published data. Only the reflections having $d > 1.2$ Å have been considered to show that the method needs few reflections and that resolution can be modest.

The first test concerns the steroid equilin (a 20-atom molecule, ignoring hydrogen atoms) with restricted conformational freedom, the second test considers sucrose, a 23-atom molecule with high conformational freedom. The latter was already considered in the first publication describing the method; the analysis has been repeated however in more challenging conditions.

All computations were done using the program TRY and assuming fixed bond-lengths in both cases, also fixed bond-angles in the case of equilin. The computing scheme is the simplest one: no hydrogen atoms, isotropic thermal vibration (an unique B_{iso}), unitary weight factors. In both cases the EA procedure consists of three phases: i) formation of the initial population, ii) improvement of solutions looking at adjacent points, iii) breeding by cross-over and mutation. The reliability of the resulting structures has been tested performing some cycles of full-matrix least-square refinement optimizing the same i.c. considered in the EA phase. The resulting R indices are, of course, not competitive with the ones published because of the crude simplicity of the followed computation scheme and because of the reduced number of reflections. The LS convergency is anyway excellent.

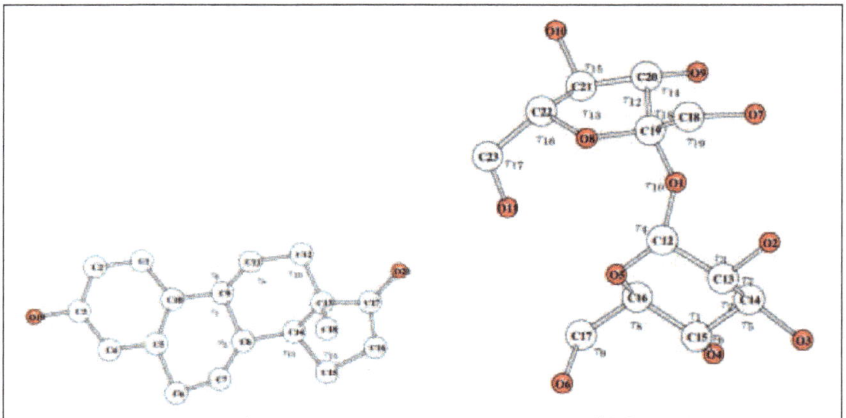

Molecular models of the two test molecules: Equilin and sucrose. Bond angles (τ_n)are shown.

	Equilin	Sucrose
Reflections considered (d > 1.2 Å)	715	589
Number of non-hydrogen atoms	20	23
Number of internal coordinates	60	68
Varied internal coordinates	15	28
Bits used in binary encoding	75	123
Random structures generated	$50X10^3$	$141X10^6$
Initial population	300	150
Lowest r_2 index after search	0.43	0.74
Lowest packing energy	-3Kcal	-11kcal
Search duration	3 min.	29 hr.
Mated structures	20	30
Resulting r_2 after breeding	0.39	0.43

References

- Libbrecht, Kenneth; Wing, Rachel (2015-09-01). The Snowflake: Winter's Frozen Artistry. Voyageur Press. ISBN 9781627887335
- Crystal-lattices-and-unit-cells, iit-jee-Solid-State: askiitians.com, Retrieved 13 May, 2020
- Wiberg, Egon; Wiberg, Nils; Holleman, Arnold Frederick (2001). Inorganic chemistry. Academic Press. p. 999. ISBN 978-0-12-352651-9
- Direct-lattice: dictionary.iucr.org, Retrieved 25 June, 2020
- Oganov, A. R.; Glass, C. W. (2006). "Crystal structure prediction using ab initio evolutionary techniques: principles and applications". Journal of Chemical Physics. 124 (24): 244704. arXiv:0911.3186. Bibcode:2006JChPh.124x47040. doi:10.1063/1.2210932. PMID 16821993

4
Crystal Systems

The space groups, lattices, point groups, crystal forms, etc. together form the crystal system. A few of its kinds are triclinic crystal system, monoclinic crystal system, tetragonal crystal system, hexagonal crystal system, cubic crystal system, orthorhombic crystal system, etc. This chapter delves into crystal system and its types for an in-depth understanding of the subject.

Crystal system is a method of classifying crystalline substances on the basis of their unit cell. There are seven unique crystal systems. The simplest and most symmetric, the cubic (or isometric) system, has the symmetry of a cube. The other six systems, in order of decreasing symmetry, are hexagonal, tetragonal, rhombohedral (also known as trigonal), orthorhombic, monoclinic and triclinic.

Triclinic Crystal System

An example of the triclinic crystals, microcline.

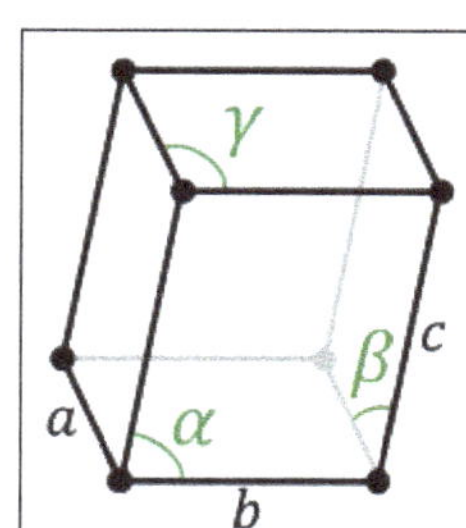

Triclinic ($a \neq b \neq c$ and $\alpha \neq \beta \neq \gamma$).

In crystallography, the triclinic (or anorthic) crystal system is one of the 7 crystal systems. A crystal system is described by three basis vectors. In the triclinic system, the crystal is described by vectors of unequal length, as in the orthorhombic system. In addition, the angles between these vectors must all be different and may not include 90°.

The triclinic lattice is the least symmetric of the 14 three-dimensional Bravais lattices. It has (itself) the minimum symmetry all lattices have: points of inversion at each lattice point and at 7 more points for each lattice point: at the midpoints of the edges and the faces, and at the center points. It is the only lattice type that itself has no mirror planes.

Crystal Classes

The triclinic crystal system class names, examples, Schönflies notation, Hermann-Mauguin notation, point groups, International Tables for Crystallography space group number, orbifold, type, and space groups are listed in the table below. There are a total of 2 space groups.

#	Point group					Type	Example	Space group
	Name	Schönflies	Intl	orbifold	Coxeter			
1	Pedial	C_1	1	11	$[\,]^+$	Enantiomorphic polar	Tantite	P1
2	Pinacoidal	C_i (S_2)	1	1×	$[2^+,2^+]$	Centrosymmetric	Wollastonite	P1

With each only one space group is associated. Pinacoidal is also known as triclinic normal. Pedial is also triclinic hemihedral.

Mineral examples include plagioclase, microcline, rhodonite, turquoise, wollastonite and amblygonite, all in triclinic normal (1).

Monoclinic Crystal System

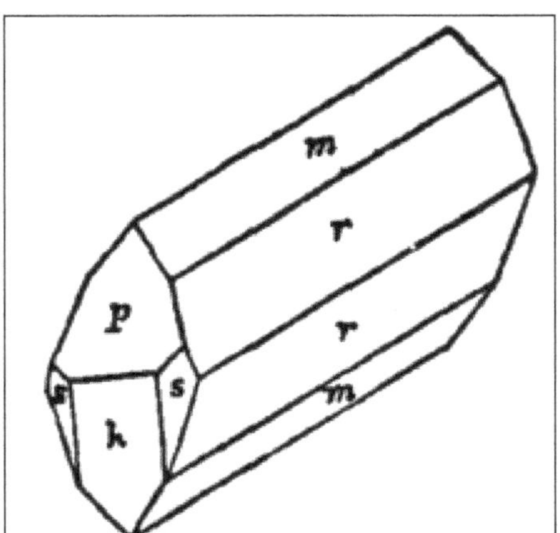

Monoclinic crystal.

In crystallography, the monoclinic crystal system is one of the 7 crystal systems. A crystal system is described by three vectors. In the monoclinic system, the crystal is described by vectors of unequal lengths, as in the orthorhombic system. They form a rectangular prism with a parallelogram as its base. Hence two vectors are perpendicular (meet at right angles), while the third vector meets the other two at an angle other than 90°.

An example of the monoclinic crystals, orthoclase.

Bravais Lattices

Two-dimensional

There is only one monoclinic Bravais lattice in two dimensions: the oblique lattice.

Three-dimensional

Two monoclinic Bravais lattices exist: the primitive monoclinic and the base-centered monoclinic lattices.

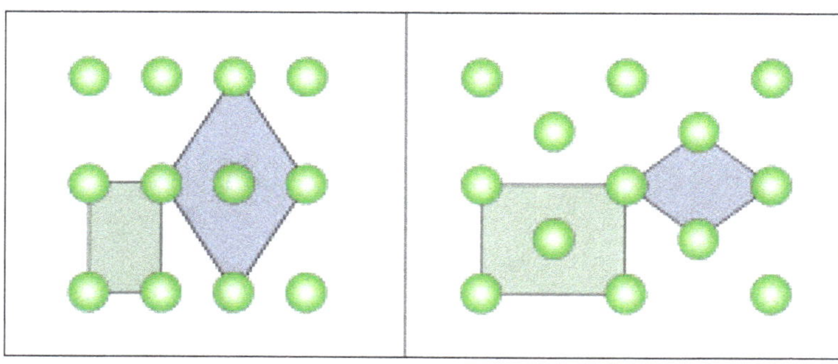

Rectangular vs rhombic unit cells for the 2D base layers of the monoclinic lattice. The two lattices swap in centering type when the axis setting is changed.

Bravais lattice	Primitive monoclinic	Base-centered monoclinic
Pearson symbol	mP	mS
Standard unit cell	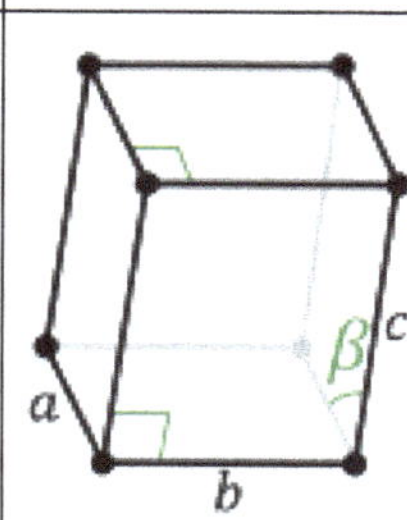	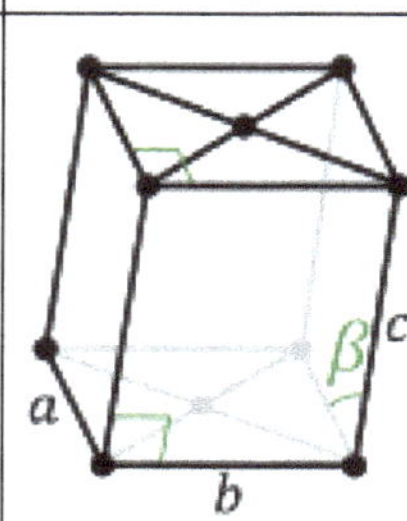

Oblique rhombic prism unit cell	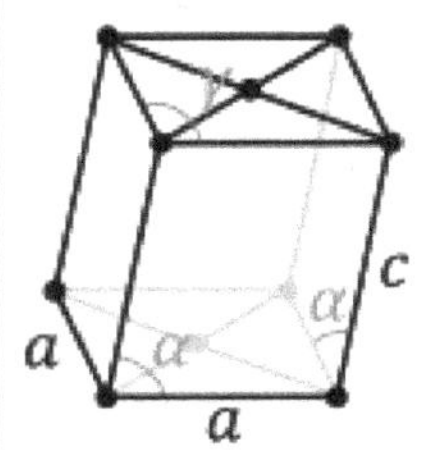	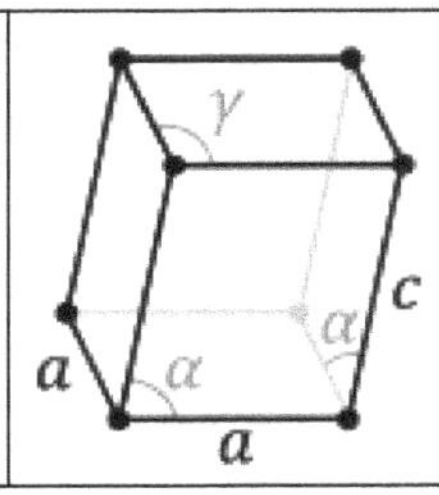

In the monoclinic system there is a rarely used second choice of crystal axes that results in a unit cell with the shape of an oblique rhombic prism; it can be constructed because the rectangular two-dimensional base layer can also be described with rhombic axes. In this axis setting, the primitive and base-centered lattices swap in centering type.

Crystal Classes

The table below organizes the space groups of the monoclinic crystal system by crystal class. It lists the International Tables for Crystallography space group numbers, followed by the crystal class name, its point group in Schoenflies notation, Hermann–Mauguin (international) notation, orbifold notation, and Coxeter notation, type descriptors, mineral examples, and the notation for the space groups.

#	Point group					Type	Example	Space groups	
	Name	Schön.	Intl	Orb.	Cox.			Primitive	Base-centered
3–5	Sphenoidal	C_2	2	22	$^+$	enantiomorphic polar	halotrichite	P2, $P2_1$	C2
6–9	Domatic	C_s (C_{1h})	m	*11	[ ]	polar	hilgardite	Pm, Pc	Cm, Cc
10–12	Prismatic	C_{2h}	2/m	2*	[2,2$^+$]	centrosymmetric	gypsum	P2/m, $P2_1/m$	C2/m
13–15								P2/c, $P2_1/c$	C2/c

Sphenoidal is also monoclinic hemimorphic; Domatic is also monoclinic hemihedral; Prismatic is also monoclinic normal.

The three monoclinic hemimorphic space groups are as follows:

- A prism with as cross-section wallpaper group p2.
- Ditto with screw axes instead of axes.
- Ditto with screw axes as well as axes, parallel, in between; in this case an additional translation vector is one half of a translation vector in the base plane plus one half of a perpendicular vector between the base planes.

The four monoclinic hemihedral space groups include:

- Those with pure reflection at the base of the prism and halfway.

- Those with glide planes instead of pure reflection planes; the glide is one half of a translation vector in the base plane.
- Those with both in between each other; in this case an additional translation vector is this glide plus one half of a perpendicular vector between the base planes.

Orthorhombic Crystal System

In crystallography, the orthorhombic crystal system is one of the 7 crystal systems. Orthorhombic lattices result from stretching a cubic lattice along two of its orthogonal pairs by two different factors, resulting in a rectangular prism with a rectangular base (a by b) and height (c), such that a, b, and c are distinct. All three bases intersect at 90° angles, so the three lattice vectors remain mutually orthogonal.

Bravais Lattices

Two-dimensional

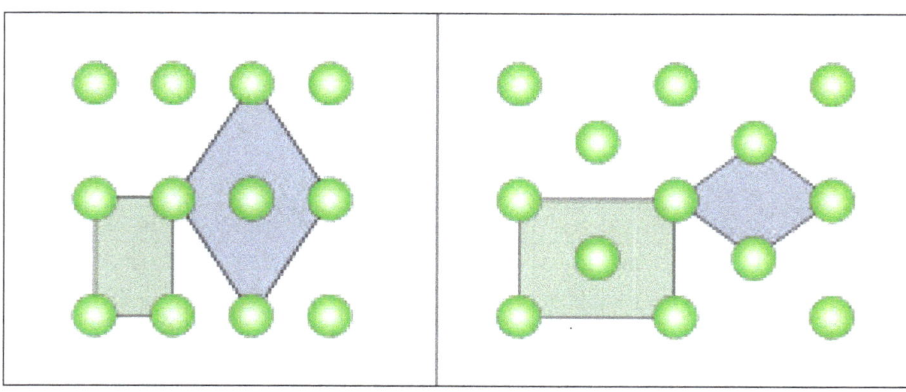

Rectangular vs rhombic unit cells for the 2D orthorhombic lattices. The swapping of centering type when the unit cell is changed also applies for the 3D orthorhombic lattices.

In two dimensions there are two orthorhombic Bravais lattices: primitive rectangular and centered rectangular. The primitive rectangular lattice can also be described by a centered rhombic unit cell, while the centered rectangular lattice can also be described by a primitive rhombic unit cell.

Three-dimensional

In three dimensions, there are four orthorhombic Bravais lattices: primitive orthorhombic, base-centered orthorhombic, body-centered orthorhombic, and face-centered orthorhombic.

Bravais lattice	Primitive orthorhombic	Base-centered orthorhombic	Body-centered orthorhombic	Face-centered orthorhombic
Pearson symbol	oP	oS	oI	oF
Standard unit cell	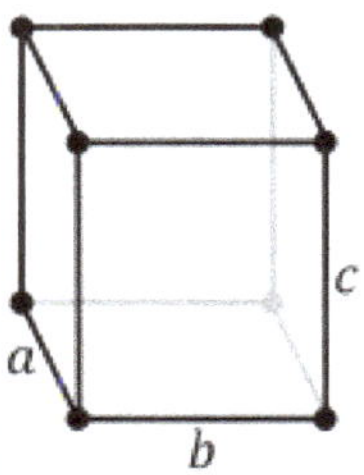	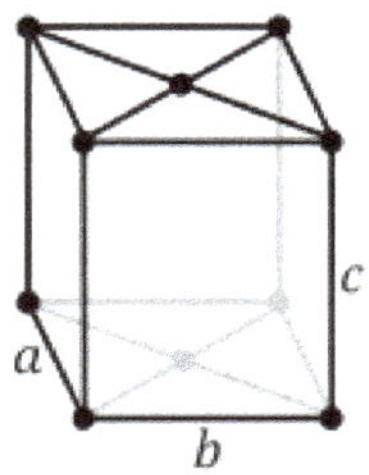	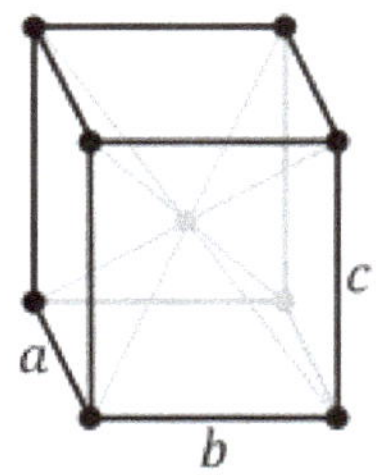	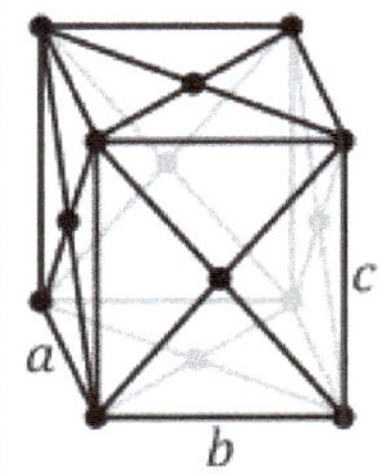
Right rhombic prism unit cell	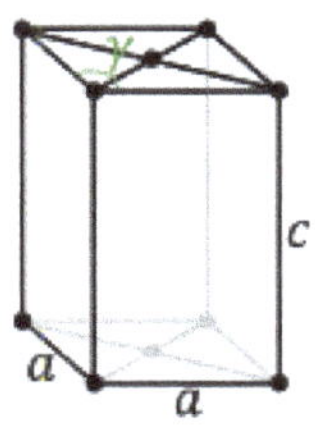	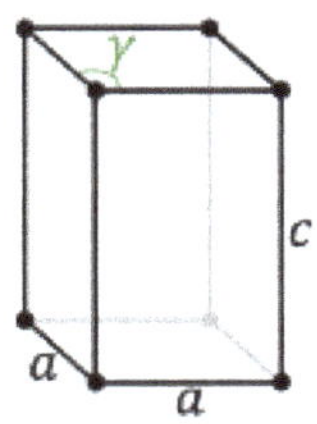	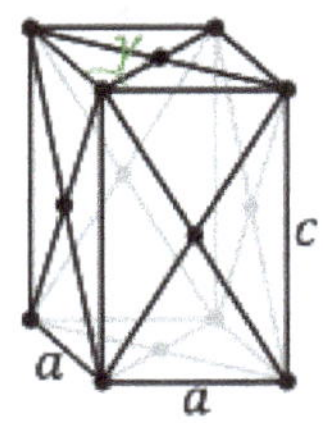	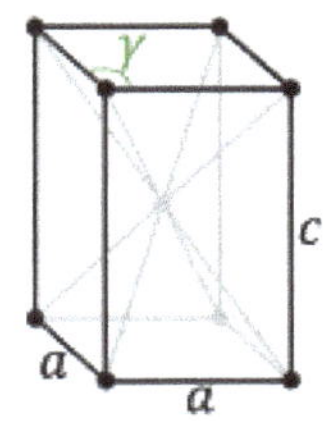

In the orthorhombic system there is a rarely used second choice of crystal axes that results in a unit cell with the shape of a right rhombic prism; it can be constructed because the rectangular two-dimensional base layer can also be described with rhombic axes. In this axis setting, the primitive and base-centered lattices swap in centering type, while the same thing happens with the body-centered and face-centered lattices. Note that the length *a* in the lower row is not the same as in the upper row, as can be seen in the figure in the section on two-dimensional lattices. For the first and third column above, *a* of the second row equals $\sqrt{a^2+b^2}$ of the first row, and for the second and fourth column it equals half of this.

Crystal Classes

The *orthorhombic crystal system* class names, examples, Schönflies notation, Hermann-Mauguin notation, point groups, International Tables for Crystallography space group number, orbifold notation, type, and space groups are listed in the table below.

№	Point group					Type	Example
	Name	Schön.	Intl	Orb.	Cox.		
16–24	Rhombic disphenoidal	D_2 (V)	222	222	$[2,2]^+$	Enantiomorphic	Epsomite
25–46	Rhombic pyramidal	C_{2v}	mm2	*22		Polar	Hemimorphite, bertrandite
47–74	Rhombic dipyramidal	D_{2h} (V_h)	mmm	*222	[2,2]	Centrosymmetric	Olivine, aragonite, marcasite
Space groups							
Primitive				Base-centered		Face-centered	Body-centered

P222, P2221, P21212, P212121	$C222_1$, C222	F222	I222, $I2_12_12_1$
Pmm2, Pmc21, Pcc2, Pma2, Pca21, Pnc2, Pmn21, Pba2, Pna21, Pnn2	Cmm2, $Cmc2_1$, Ccc2 Amm2, Aem2, Ama2, Aea2	Fmm2, Fdd2	Imm2, Iba2, Ima2
Pmmm, Pnnn, Pccm, Pban, Pmma, Pnna, Pmna, Pcca, Pbam, Pccn, Pbcm, Pnnm, Pmmn, Pbcn, Pbca, Pnma	Cmcm, Cmca, Cmmm, Cccm, Cmme, Ccce	Fmmm, Fddd	Immm, Ibam, Ibca, Imma

Tetragonal Crystal System

In crystallography, the tetragonal crystal system is one of the 7 crystal systems. Tetragonal crystal lattices result from stretching a cubic lattice along one of its lattice vectors, so that the cube becomes a rectangular prism with a square base (*a* by *a*) and height (*c*, which is different from *a*).

An example of the tetragonal crystals, wulfenite.

Bravais Lattices

Two-dimensional

There is only one tetragonal Bravais lattice in two dimensions: the square lattice.

Three-dimensional

There are two tetragonal Bravais lattices: the simple tetragonal (from stretching the simple-cubic lattice) and the centered tetragonal (from stretching either the face-centered or the body-centered cubic lattice). One might suppose stretching face-centered cubic would result in face-centered tetragonal, but the face-centered tetragonal is equivalent to the body-centered tetragonal, BCT (with a smaller lattice spacing). BCT is considered more fundamental, and therefore this is the standard terminology.

Bravais lattice	Primitive tetragonal	Body-centered tetragonal
Pearson symbol	tP	tI
Unit cell	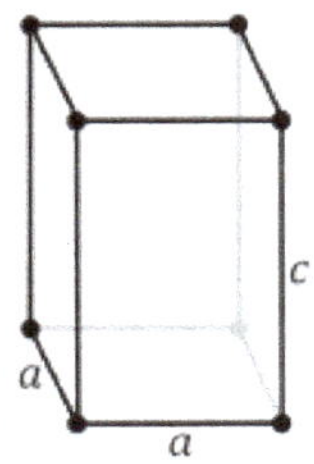	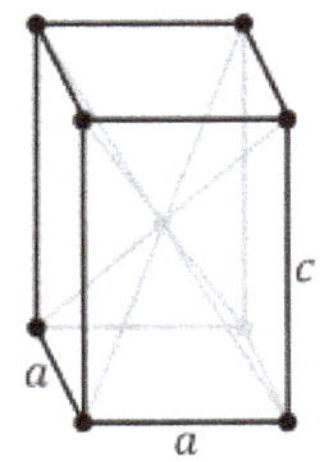

Hexagonal Crystal System

Crystal system	Trigonal		Hexagonal
Lattice system	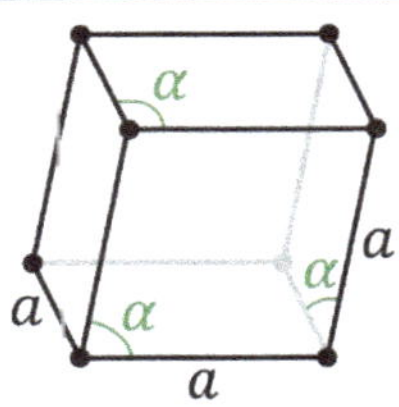 Rhombohedral	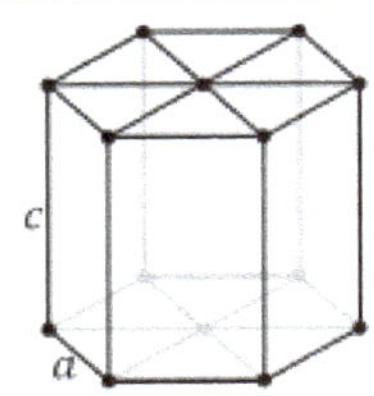 Hexagonal	
Example	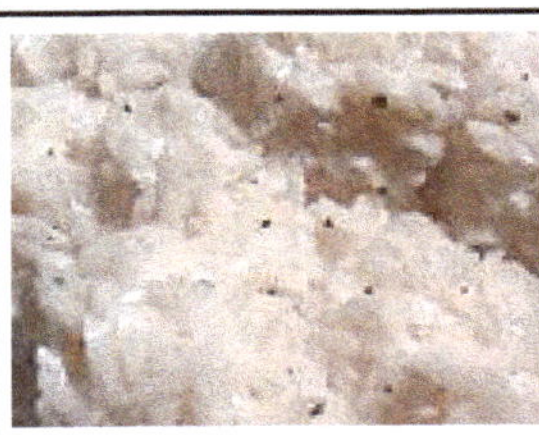 Dolomite	 Cinnabar	 Beryl

In crystallography, the hexagonal crystal family is one of the 6 crystal families, which includes 2 crystal systems (hexagonal and trigonal) and 2 lattice systems (hexagonal and rhombohedral).

The hexagonal crystal family consists of the 12 point groups such that at least one of their space groups has the hexagonal lattice as underlying lattice, and is the union of the hexagonal crystal system and the trigonal crystal system. There are 52 space groups associated with it, which are exactly those whose Bravais lattice is either hexagonal or rhombohedral.

Lattice Systems

The hexagonal crystal family consists of two lattice systems: hexagonal and rhombohedral. Each lattice system consists of one Bravais lattice.

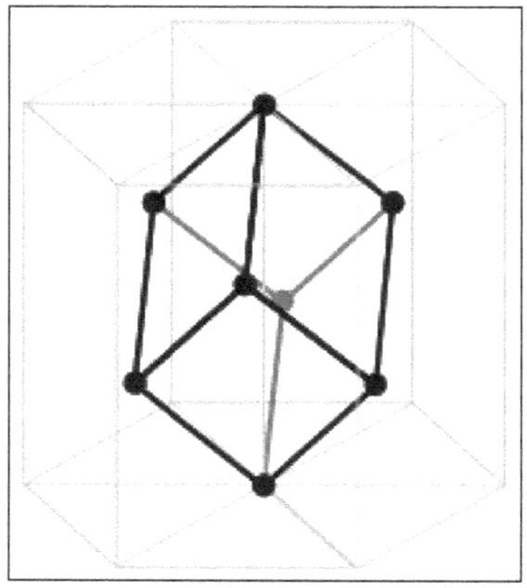

Relation between the two settings for the rhombohedral lattice.

Hexagonal crystal family		
Bravais lattice	Hexagonal	Rhombohedral
Pearson symbol	hP	hR
Hexagonal unit cell	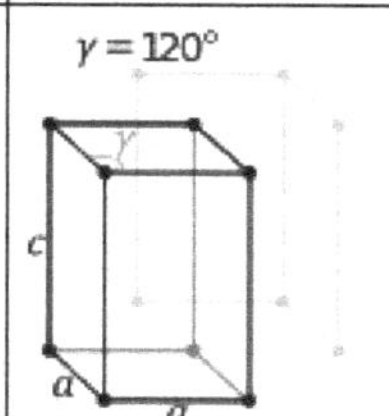	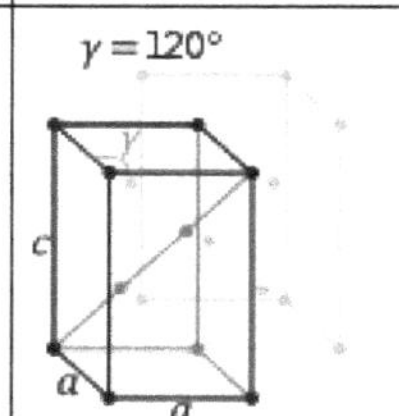
Rhombohedral unit cell	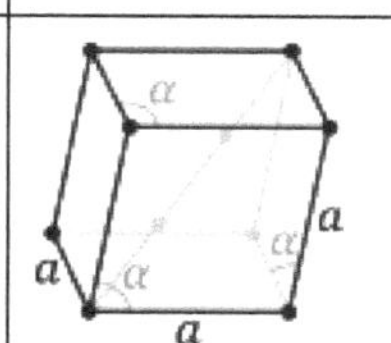	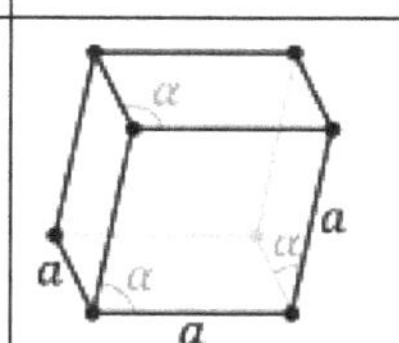

In the hexagonal family, the crystal is conventionally described by a right rhombic prism unit cell with two equal axes (a by a), an included angle of 120° (γ) and a height (c, which can be different from a) perpendicular to the two base axes.

The hexagonal unit cell for the rhombohedral Bravais lattice is the R-centered cell, consisting of two additional lattice points which occupy one body diagonal of the unit cell. There are two ways to do this, which can be thought of as two notations which represent the same structure. In the usual so-called obverse setting, the additional lattice points are at coordinates ($\frac{2}{3}$, $\frac{1}{3}$, $\frac{1}{3}$) and ($\frac{1}{3}$, $\frac{2}{3}$, $\frac{2}{3}$), whereas in the alternative reverse setting they are at the coordinates ($\frac{1}{3}$,$\frac{2}{3}$,$\frac{1}{3}$) and ($\frac{2}{3}$,$\frac{1}{3}$,$\frac{2}{3}$). In either case, there are 3 lattice points per unit cell in total and the lattice is non-primitive.

The Bravais lattices in the hexagonal crystal family can also be described by rhombohedral axes. The unit cell is a rhombohedron (which gives the name for the rhombohedral lattice). This is a unit cell with parameters $a = b = c$; $\alpha = \beta = \gamma \neq 90°$. In practice, the hexagonal description is more commonly used because it is easier to deal with a coordinate system with two 90° angles. However, the rhombohedral axes are often shown (for the rhombohedral lattice) in textbooks because this cell reveals $\bar{3}m$ symmetry of crystal lattice.

The rhombohedral unit cell for the hexagonal Bravais lattice is the D-centered cell, consisting of two additional lattice points which occupy one body diagonal of the unit cell with coordinates ($\frac{1}{3}$, $\frac{1}{3}$, $\frac{1}{3}$) and ($\frac{2}{3}$, $\frac{2}{3}$, $\frac{2}{3}$). However, such a description is rarely used.

Crystal Systems

Crystal system	Required symmetries of point group	Point groups	Space groups	Lattice system
Trigonal	1 threefold axis of rotation	5	7	Rhombohedral
			18	Hexagonal
Hexagonal	1 sixfold axis of rotation	7	27	

The hexagonal crystal family consists of two crystal systems: trigonal and hexagonal. A crystal system is a set of point groups in which the point groups themselves and their corresponding space groups are assigned to a lattice system.

The trigonal crystal system consists of the 5 point groups that have a single three-fold rotation axis, which includes space groups 143 to 167. These 5 point groups have 7 corresponding space groups (denoted by R) assigned to the rhombohedral lattice system and 18 corresponding space groups (denoted by P) assigned to the hexagonal lattice system.

The hexagonal crystal system consists of the 7 point groups that have a single six-fold rotation axis. These 7 point groups have 27 space groups (168 to 194), all of which are assigned to the hexagonal lattice system. Graphite is an example of a crystal that crystallizes in the hexagonal crystal system.

Hexagonal Close Packed

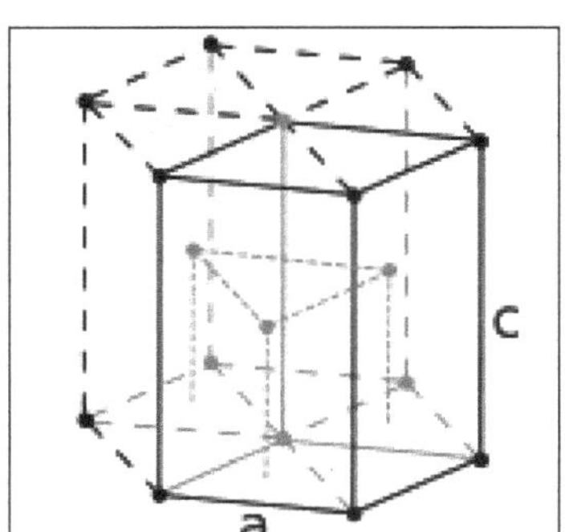

Hexagonal close packed (hcp) unit cell.

Hexagonal close packed (hcp) is one of the two simple types of atomic packing with the highest density, the other being the face centered cubic (fcc). However, unlike the fcc, it is not a Bravais lattice as there are two nonequivalent sets of lattice points. Instead, it can be constructed from the hexagonal Bravais lattice by using a two atom motif (the additional atom at about ($\frac{2}{3}$,$\frac{1}{3}$,$\frac{1}{2}$)) associated with each lattice point.

Quartz

Quartz mineral embedded in limestone (top right of the sample), easily identifiable by its hexagonal form.

Quartz is a crystal that belongs to the hexagonal lattice system but exists in two polymorphs that are in two different crystal systems. The crystal structures of α-quartz are described by two of the 18 space groups (152 and 154) associated with the trigonal crystal system, while the crystal structures of β-quartz are described by two of the 27 space groups (180 and 181) associated with the hexagonal crystal system.

Rhombohedral Lattice Angle

The lattice angles and the lengths of the lattice vectors are all the same for both the cubic and rhombohedral lattice systems. The lattice angles for simple cubic, face-centered cubic, and body-centered cubic lattices are $\pi/2$ radians, $\pi/3$ radians, and arccos(-1/3) radians, respectively. A rhombohedral lattice will result from lattice angles other than these.

Cubic Crystal System

In crystallography, the cubic (or isometric) crystal system is a crystal system where the unit cell is in the shape of a cube. This is one of the most common and simplest shapes found in crystals and minerals.

A rock containing three crystals of pyrite (FeS_2). The crystal structure of pyrite is primitive cubic, and this is reflected in the cubic symmetry of its natural crystal facets.

There are three main varieties of these crystals:

- Primitive cubic (abbreviated cP and alternatively called simple cubic).
- Body-centered cubic (abbreviated cI or bcc).
- Face-centered cubic (abbreviated cF or fcc, and alternatively called cubic close-packed or ccp).

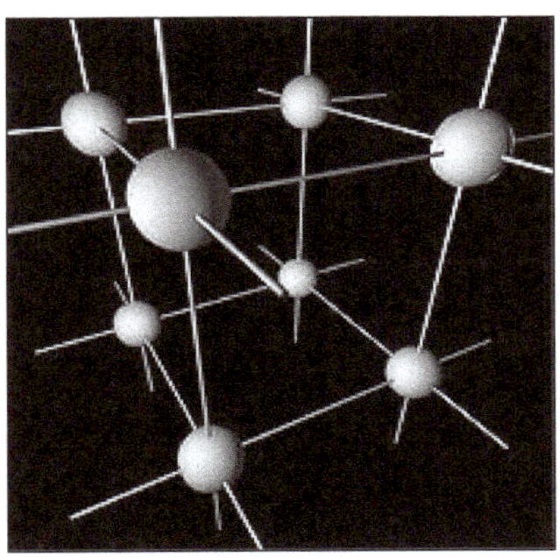

A network model of a primitive cubic system.

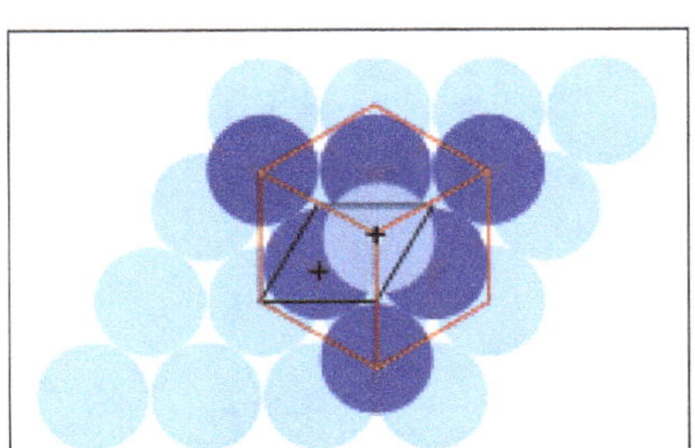

The primitive and cubic close-packed (also known as face-centered cubic) unit cells.

Each is subdivided into other variants listed below. Note that although the *unit cell* in these crystals is conventionally taken to be a cube, the primitive unit cell often is not.

Bravais Lattices

The three Bravais lattices in the cubic crystal system are:

Bravais lattice	Primitive cubic	Body-centered cubic	Face-centered cubic
Pearson symbol	cP	cI	cF
Unit cell	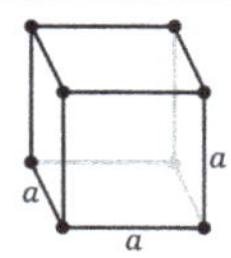	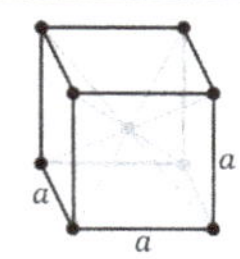	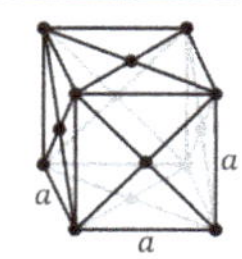

The primitive cubic system (*cP*) consists of one lattice point on each corner of the cube. Each atom at a lattice point is then shared equally between eight adjacent cubes, and the unit cell therefore contains in total one atom ($\frac{1}{8} \times 8$). The body-centered cubic system (*cI*) has one lattice point in the center of the unit cell in addition to the eight corner points. It has a net total of 2 lattice points per unit cell ($\frac{1}{8} \times 8 + 1$). The face-centered cubic system (*cF*) has lattice points on the faces of the cube, that each gives exactly one half contribution, in addition to the corner lattice points, giving a total of 4 lattice points per unit cell ($\frac{1}{8} \times 8$ from the corners plus $\frac{1}{2} \times 6$ from the faces). Each sphere in a cF lattice has coordination number 12. Coordination number is the number of nearest neighbours of a central atom in the structure.

The face-centered cubic system is closely related to the hexagonal close packed (hcp) system, where two systems differ only in the relative placements of their hexagonal layers. The plane of a face-centered cubic system is a hexagonal grid.

Attempting to create a C-centered cubic crystal system (i.e., putting an extra lattice point in the center of each horizontal face) would result in a simple tetragonal Bravais lattice.

Voids in the Unit Cell

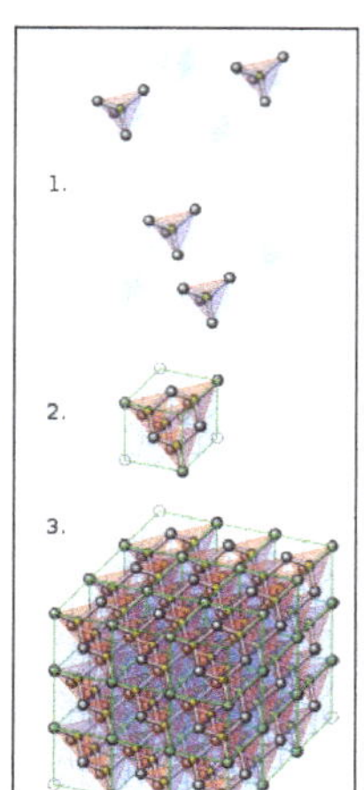

Visualisation of a diamond cubic unit cell: 1. Components of a unit cell, 2. One unit cell, 3. A lattice of 3 x 3 x 3 unit cells.

A simple cubic unit cell has a single cubic void in the center.

A body-centered cubic unit cell has six octahedral voids located at the center of each face of the unit cell, and twelve further ones located at the midpoint of each edge of the same cell, for a total of six net octahedral voids. Additionally, there are 24 tetrahedral voids located in a square spacing around each octahedral void, for a total of twelve net tetrahedral voids. These tetrahedral voids are not local maxima and are not technically voids, but they do occasionally appear in multi-atom unit cells.

A face-centered cubic unit cell has eight tetrahedral voids located midway between each corner and the center of the unit cell, for a total of eight net tetrahedral voids. Additionally, there are twelve octahedral voids located at the midpoints of the edges of the unit cell as well as one octahedral hole in the very center of the cell, for a total of four net octahedral voids.

One important characteristic of a crystalline structure is its atomic packing factor. This is calculated by assuming that all the atoms are identical spheres, with a radius large enough that each sphere abuts on the next. The atomic packing factor is the proportion of space filled by these spheres.

Assuming one atom per lattice point, in a primitive cubic lattice with cube side length a, the sphere radius would be $a/2$ and the atomic packing factor turns out to be about 0.524 (which is quite low). Similarly, in a *bcc* lattice, the atomic packing factor is 0.680, and in *fcc* it is 0.740. The *fcc* value is the highest theoretically possible value for any lattice, although there are other lattices which also achieve the same value, such as hexagonal close packed (*hcp*) and one version of tetrahedral *bcc*.

As a rule, since atoms in a solid attract each other, the more tightly packed arrangements of atoms tend to be more common. (Loosely packed arrangements do occur, though, for example if the orbital hybridization demands certain bond angles.) Accordingly, the primitive cubic structure, with especially low atomic packing factor, is rare in nature, but is found in polonium. The *bcc* and *fcc*, with their higher densities, are both quite common in nature. Examples of *bcc* include iron, chromium, tungsten, and niobium. Examples of *fcc* include aluminium, copper, gold and silver.

Multi-element Compounds

Compounds that consist of more than one element (e.g. binary compounds) often have crystal structures based on a cubic crystal system. Some of the more common ones are listed here.

Caesium Chloride Structure

The space group of the caesium chloride (CsCl) structure is called $Pm\bar{3}m$ (in Hermann–Mauguin notation), or “221” (in the International Tables for Crystallography). The Strukturbericht designation is “B2”.

One structure is the "interpenetrating primitive cubic" structure, also called the "caesium chloride" structure. Each of the two atom types forms a separate primitive cubic lattice, with an atom of one type at the center of each cube of the other type. In the unit cell of CsCl, each ion is at the center of a cube of ions of the opposite kind, so the co - ordination number is eight. Altogether, the arrangement of atoms is the same as body-centered cubic, but with alternating types of atoms at the different lattice sites. Alternately, one could view this lattice as a simple cubic structure with a secondary atom in its cubic void.

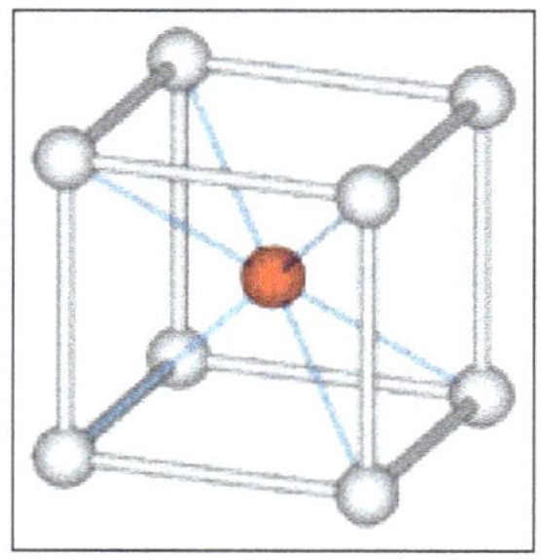

A caesium chloride unit cell. The two colors of spheres represent the two types of atoms.

In addition to caesium chloride itself, the structure also appears in certain other alkali halides when prepared at low temperatures or high pressures. Generally, this structure is more likely to be formed from two elements whose ions are of roughly the same size (for example, ionic radius of Cs^+ = 167 pm, and Cl^- = 181 pm).

The coordination number of each atom in the structure is 8: the central cation is coordinated to 8 anions on the corners of a cube as shown, and similarly, the central anion is coordinated to 8 cations on the corners of a cube.

Other compounds showing caesium chloride like structure are CsBr, CsI, high-temp RbCl, AlCo, AgZn, BeCu, MgCe, RuAl and SrTl.

Rock-salt Structure

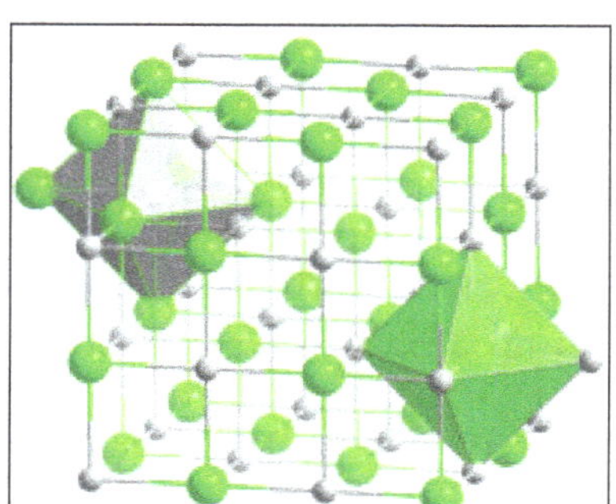

The rock-salt crystal structure. Each atom has six nearest neighbors, with octahedral geometry.

The space group of the rock-salt (NaCl) structure is called $Fm\bar{3}m$ (in Hermann–Mauguin notation), or "225" (in the International Tables for Crystallography). The Strukturbericht designation is "B1".

In the rock-salt or sodium chloride (halite) structure, each of the two atom types forms a separate face-centered cubic lattice, with the two lattices interpenetrating so as to form a 3D checkerboard pattern. Alternately, one could view this structure as a face-centered cubic structure with secondary atoms in its octahedral holes.

Examples of compounds with this structure include sodium chloride itself, along with almost all other alkali halides, and "many divalent metal oxides, sulfides, selenides, and tellurides". More generally, this structure is more likely to be formed if the cation is somewhat smaller than the anion (a cation/anion radius ratio of 0.414 to 0.732).

The coordination number of each atom in this structure is 6: each cation is coordinated to 6 anions at the vertices of an octahedron, and similarly, each anion is coordinated to 6 cations at the vertices of an octahedron.

The interatomic distance (distance between cation and anion, or half the unit cell length a) in some rock-salt-structure crystals are: 2.3 Å (2.3×10^{-10} m) for NaF, 2.8 Å for NaCl, and 3.2 Å for SnTe.

Other compounds showing rock salt like structure are LiF, LiCl, LiBr, LiI, NaF, NaBr, NaI, KF, KCl, KBr, KI, RbF, RbCl, RbBr, RbI, CsF, MgO, PbS, AgF, AgCl, AgBr and ScN.

Zincblende Structure

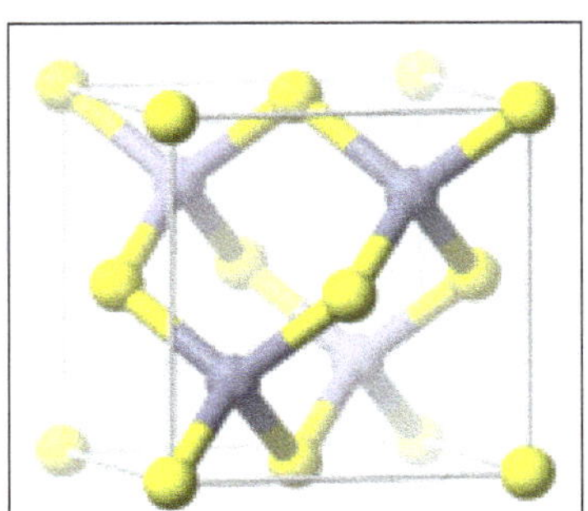

A zincblende unit cell.

The space group of the Zincblende structure is called $F\bar{4}3m$ (in Hermann–Mauguin notation), or 216. The Strukturbericht designation is "B3".

The Zincblende structure (also written "zinc blende") is named after the mineral zincblende (sphalerite), one form of zinc sulfide (β-ZnS). As in the rock-salt structure, the two atom types form two interpenetrating face-centered cubic lattices. However, it differs from rock-salt structure in how the two lattices are positioned relative to one another. The zincblende structure has tetrahedral coordination: Each atom's nearest neighbors consist of four atoms of the opposite type, positioned like the four vertices of a regular tetrahedron. Altogether, the arrangement of atoms in zincblende structure is the same as diamond cubic structure, but with alternating types of atoms at the different lattice sites.

Examples of compounds with this structure include zincblende itself, lead(II) nitrate, many compound semiconductors (such as gallium arsenide and cadmium telluride), and a wide array of other binary compounds.

Other compounds showing zinc blende-like structure are α-AgI, β-BN, diamond, CuBr, β-CdS, BP and BAs.

Weaire–Phelan Structure

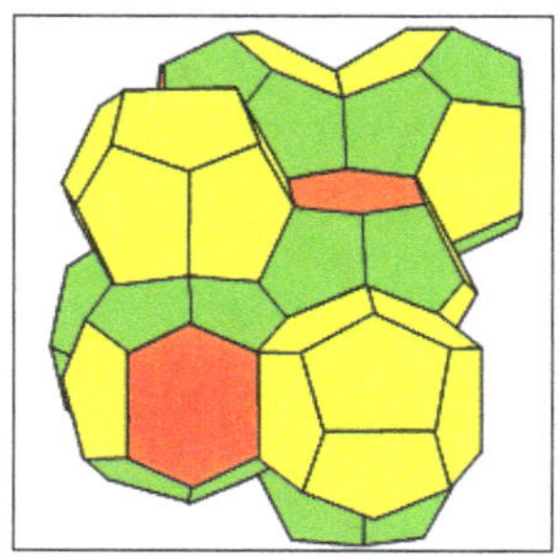

Weaire–Phelan structure.

The Weaire–Phelan structure has Pm3n (223) symmetry.

It has 3 orientations of stacked tetradecahedrons with pyritohedral cells in the gaps. It is found as a crystal structure in chemistry where it is usually known as the "Type I clathrate structure". Gas hydrates formed by methane, propane, and carbon dioxide at low temperatures have a structure in which water molecules lie at the nodes of the Weaire–Phelan structure and are hydrogen bonded together, and the larger gas molecules are trapped in the polyhedral cages.

Space Group

In crystallography, Space group is any of the ways in which the orientation of a crystal can be changed without seeming to change the position of its atoms. These changes may involve displacement of the whole structure along a crystallographic axis (translation), as well as the point group operations of rotation about an axis, reflection in a plane, inversion about a centre, or sequential rotary inversion. As demonstrated in the 1890s, only 230 distinct combinations of these changes are possible; these 230 combinations define the 230 space groups. A crystal can be assigned to one of these groups after the arrangement of its atoms is studied, as by X-ray crystallography, thereby providing a definitive way of categorizing the inherent symmetry of the crystal.

Bravais Lattices

In crystallography, a Bravais lattice, named after Auguste Bravais is an infinite array of discrete points generated by a set of discrete translation operations described in three dimensional space by:

$$\mathbf{R} = n_1\mathbf{a}_1 + n_2\mathbf{a}_2 + n_3\mathbf{a}_3$$

where n_i are any integers and $\mathbf{a}_i$ are *primitive vectors* which lie in different directions (not necessarily mutually perpendicular) and span the lattice. This discrete set of vectors must be closed under vector addition and subtraction. For any choice of position vector **R**, the lattice looks exactly the same.

When the discrete points are atoms, ions, or polymer strings of solid matter, the Bravais lattice concept is used to formally define a crystalline arrangement and its (finite) frontiers. A crystal is made up of a periodic arrangement of one or more atoms (the basis, or motif) repeated at each lattice point. Consequently, the crystal looks the same when viewed from any equivalent lattice point, namely those separated by the translation of one unit cell.

Two Bravais lattices are often considered equivalent if they have isomorphic symmetry groups. In this sense, there are 14 possible Bravais lattices in three-dimensional space. The 14 possible symmetry groups of Bravais lattices are 14 of the 230 space groups. In the context of the space group classification, the Bravais lattices are also called Bravais classes, Bravais arithmetic classes, or Bravais flocks.

2 Dimensions

Table: 1 – oblique (monoclinic), 2 – rectangular (orthorhombic), 3 – centered rectangular (orthorhombic), 4 – hexagonal, and 5 – square (tetragonal).

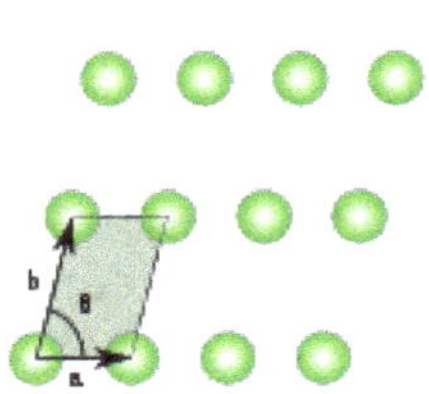 1	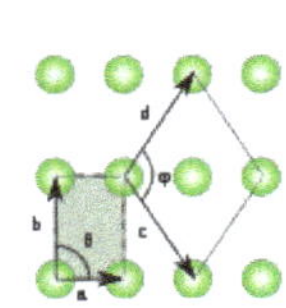 2 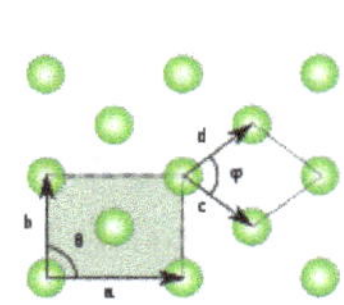 3	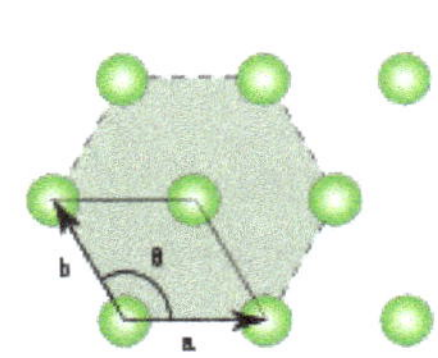 4	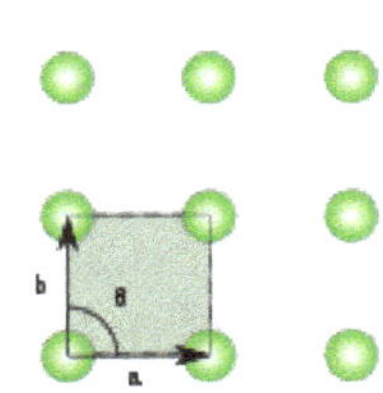 5
$\lvert a\rvert \neq \lvert b\rvert, \theta \neq 90^\circ$ m	$\lvert a\rvert \neq \lvert b\rvert, \theta = 90^\circ$ $\lvert c\rvert = \lvert d\rvert, \varphi \neq 90^\circ$ o	$\lvert a\rvert = \lvert b\rvert, \theta = 120^\circ$ h	$\lvert a\rvert = \lvert b\rvert, \theta = 90^\circ$ t

In two-dimensional space, there are 5 Bravais lattices, grouped into four crystal families.

Crystal family	Point group (Schönflies notation)	5 Bravais lattices	
		Primitive	Centered
Monoclinic	C_2	Oblique	
Orthorhombic	D_2	Rectangular	Centered rectangular
Hexagonal	D_6	Hexagonal	
Tetragonal	D_4	Square	

The unit cells are specified according to the relative lengths of the cell edges (a and b) and the angle between them (θ). The area of the unit cell can be calculated by evaluating the norm $||\mathbf{a} \times \mathbf{b}||$, where **a** and **b** are the lattice vectors. The properties of the crystal families are given below:

Crystal family	Area	Axial distances (edge lengths)	Axial angle
Monoclinic	$ab\sin\theta$	$a \neq b$	$\theta \neq 90°$
Orthorhombic	ab	$a \neq b$	$\theta = 90°$
Hexagonal	$\frac{3\sqrt{3}}{2}a^2$	$a = b$	$\theta = 120°$
Tetragonal	a^2	$a = b$	$\theta = 90°$

3 Dimensions

2×2×2 unit cells of a diamond cubic lattice.

In three-dimensional space, there are 14 Bravais lattices. These are obtained by combining one of the seven lattice systems with one of the centering types. The centering types identify the locations of the lattice points in the unit cell as follows:

- Primitive (P): Lattice points on the cell corners only (sometimes called simple).
- Base-centered (A, B, or C): Lattice points on the cell corners with one additional point at the center of each face of one pair of parallel faces of the cell (sometimes called end-centered).

- Body-centered (I): Lattice points on the cell corners, with one additional point at the center of the cell.
- Face-centered (F): Lattice points on the cell corners, with one additional point at the center of each of the faces of the cell.

Not all combinations of lattice systems and centering types are needed to describe all of the possible lattices, as it can be shown that several of these are in fact equivalent to each other. For example, the monoclinic I lattice can be described by a monoclinic C lattice by different choice of crystal axes. Similarly, all A- or B-centred lattices can be described either by a C- or P-centering.

The unit cells are specified according to the relative lengths of the cell edges (a, b, c) and the angles between them (α, β, γ). The volume of the unit cell can be calculated by evaluating the triple product $\mathbf{a} \cdot (\mathbf{b} \times \mathbf{c})$, where **a**, **b**, and **c** are the lattice vectors.

4 Dimensions

In four dimensions, there are 64 Bravais lattices. Of these, 23 are primitive and 41 are centered. Ten Bravais lattices split into enantiomorphic pairs.

Miller Indices

Miller indices is group of three numbers that indicates the orientation of a plane or set of parallel planes of atoms in a crystal. If each atom in the crystal is represented by a point and these points are connected by lines, the resulting lattice may be divided into a number of identical blocks, or unit cells; the intersecting edges of one of the unit cells defines a set of crystallographic axes, and the Miller indices are determined by the intersection of the plane with these axes. The reciprocals of these intercepts are computed, and fractions are cleared to give the three Miller indices (*hkl*). For example, a plane parallel to two axes but cutting the third axis at a length equal to one edge of a unit cell has Miller indices of (100), (010), or (001), depending upon the axis cut; and a plane cutting all three axes at lengths equal to the edges of a unit cell has Miller indices of (111). This scheme, devised by British mineralogist and crystallographer William Hallowes Miller, in 1839, has the advantage of eliminating all fractions from the notation for a plane. In the hexagonal system, which has four crystallographic axes, a similar scheme of four Bravais-Miller indices is used.

Calculating Miller Indices

The orientation of a surface or a crystal plane may be defined by considering how the plane (or indeed any parallel plane) intersects the main crystallographic axes of the solid. The application of a set of rules leads to the assignment of the Miller Indices (*hkl*), which are a set of numbers which quantify the intercepts and thus may be used to uniquely identify the plane or surface.

The following treatment of the procedure used to assign the Miller Indices is a simplified one (it may be best if you simply regard it as a "recipe") and only a cubic crystal system (one having a cubic unit cell with dimensions $a \times a \times a$) will be considered.

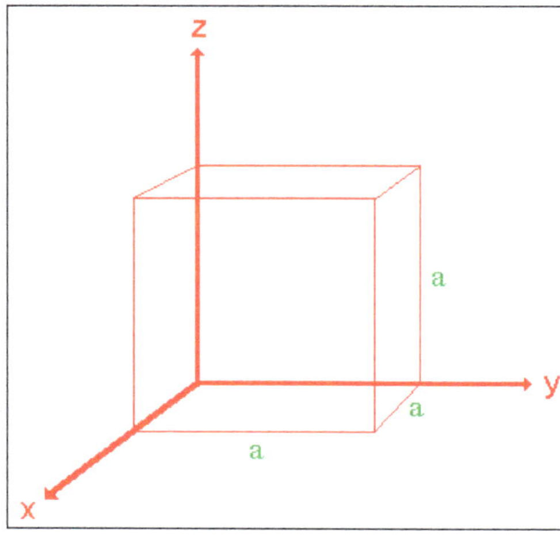

The procedure is most easily illustrated using an example so we will first consider the following surface/plane:

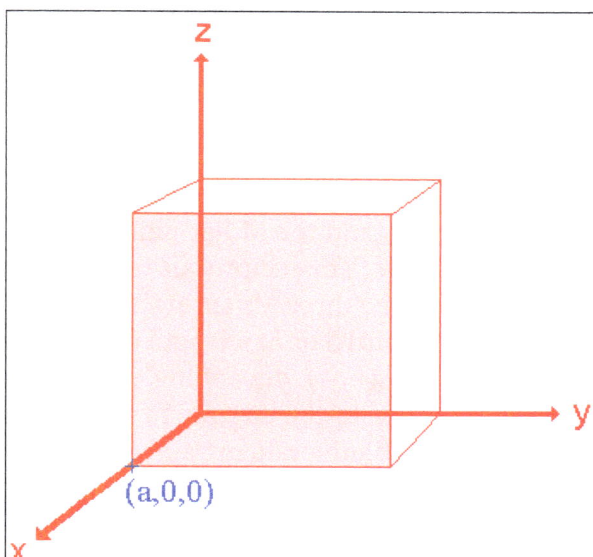

Step 1: Identify the intercepts on the x- , y- and z- axes.

In this case the intercept on the x-axis is at $x = a$ (at the point (a,0,0)), but the surface is parallel to the y- and z-axes - strictly therefore there is no intercept on these two axes but we shall consider the intercept to be at infinity (∞) for the special case where the plane is parallel to an axis. The intercepts on the x- , y- and z-axes are thus:

Intercepts: a , ∞ , ∞

Step 2: Specify the intercepts in fractional co-ordinates.

Co-ordinates are converted to fractional co-ordinates by dividing by the respective cell-dimension - for example, a point (x,y,z) in a unit cell of dimensions a x b x c has fractional co-ordinates of $(x/a , y/b , z/c)$. In the case of a cubic unit cell each co-ordinate will simply be divided by the cubic cell constant , a . This gives:

Fractional Intercepts: $\boldsymbol{a/a}$, $\infty/\boldsymbol{a}$, ∞/a i.e. **1** , ∞ , ∞

Step 3: Take the reciprocals of the fractional intercepts.

This final manipulation generates the Miller Indices which (by convention) should then be specified without being separated by any commas or other symbols. The Miller Indices are also enclosed within standard brackets when one is specifying a unique surface such as that being considered here.

The reciprocals of 1 and ∞ are 1 and 0 respectively, thus yielding.

Miller Indices: (100).

So the surface/plane illustrated is the (100) plane of the cubic crystal.

Other Examples of Miller Indices

The (110) Surface

- Intercepts: a , a , ∞
- Fractional intercepts: 1 , 1 , ∞
- Miller Indices: (110)

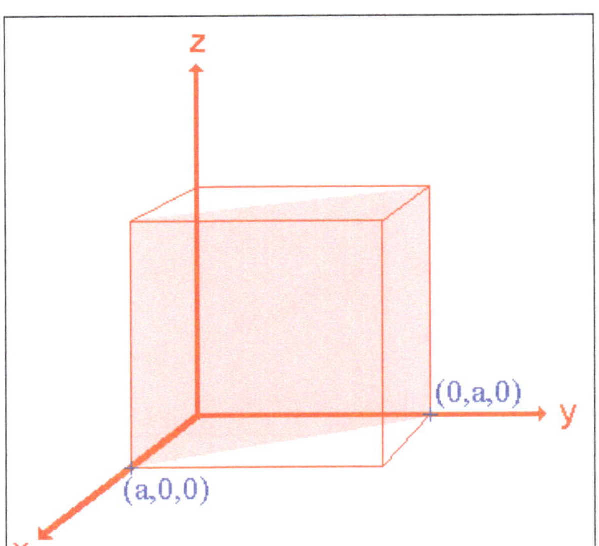

The (111) Surface

- Intercepts: a , a , a
- Fractional intercepts: 1 , 1 , 1
- Miller Indices: (111)

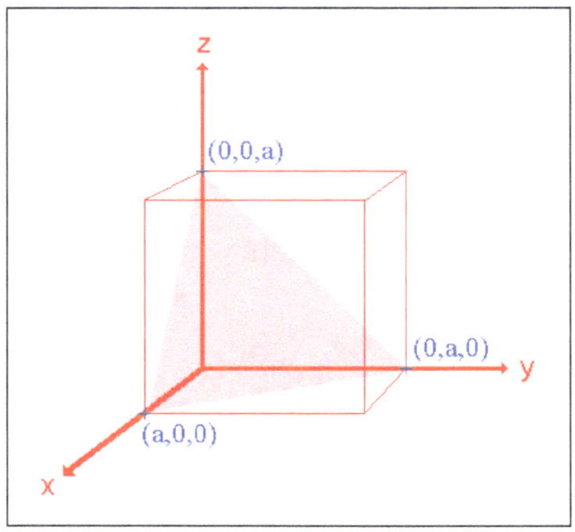

The (100), (110) and (111) surfaces considered above are the so-called low index surfaces of a cubic crystal system (the "low" refers to the Miller indices being small numbers - 0 or 1 in this case). These surfaces have a particular importance but there an infinite number of other planes that may be defined using Miller index notation.

The (210) Surface

- Intercepts: ½ a , a , ∞
- Fractional intercepts: ½ , 1 , ∞
- Miller Indices: (210)

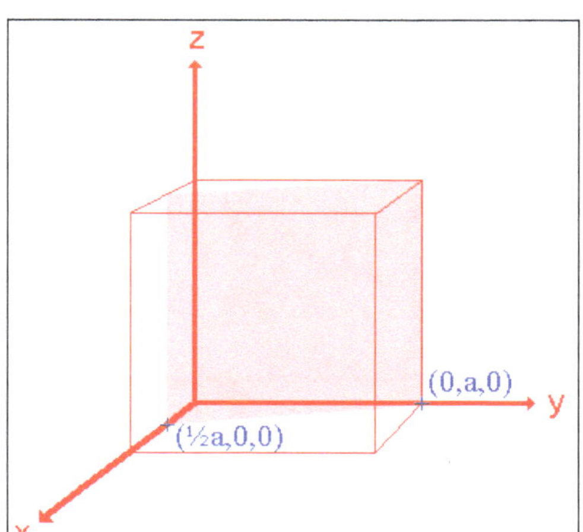

Law of Rational Indices

The law of rational indices states that the intercepts, OP, OQ, OR, of the natural faces of a crystal form with the unit-cell axes a, b, c are inversely proportional to prime integers, h, k, l. They are called the Miller indices of the face. They are usually small because the corresponding lattice planes are among the densest and have therefore a high interplanar spacing and low indices.

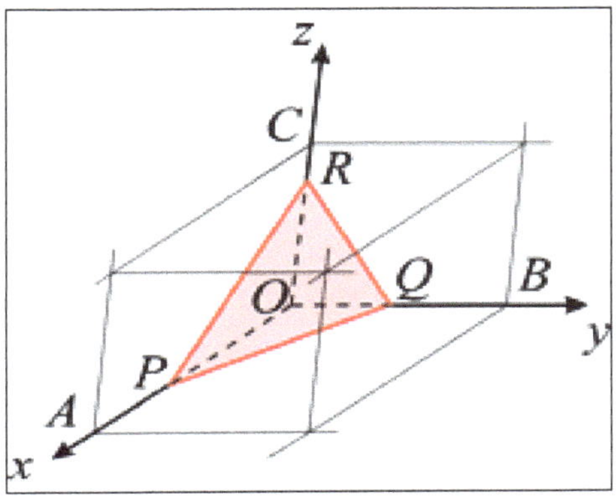

The law of rational indices was deduced by Haüy from the observation of the stacking laws required to build the natural faces of crystals by piling up elementary blocks, for instance cubes to construct the {110} faces of the rhomb-dodecahedron observed in garnets or the ½{210} faces of the pentagon-dodecahedron observed in pyrite, or rhombohedrons to construct the {21.1} (referred to an hexagonal lattice, $\{2\bar{1}0\}$ referred to a rhombohedral lattice) scalenohedron of calcite.

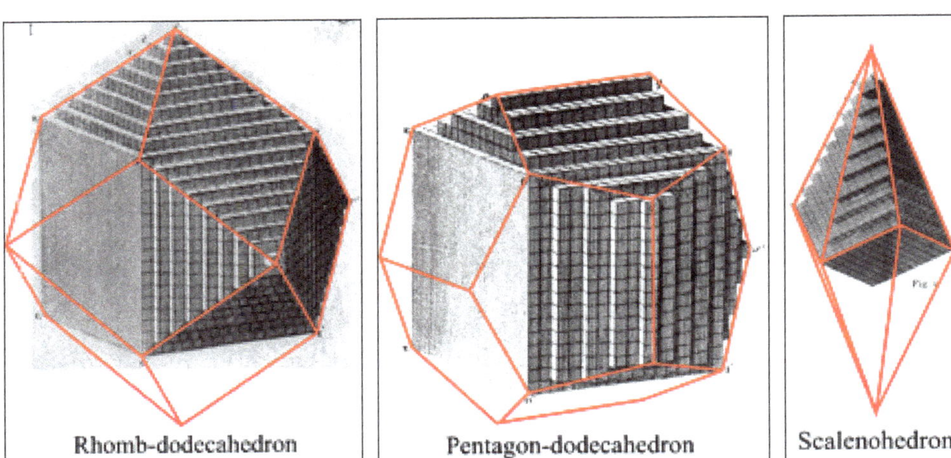

The crystal forms have been superimposed in heavy lines.

References

- Prince, E., ed. (2006). International Tables for Crystallography. International Union of Crystallography. doi:10.1107/97809553602060000001. ISBN 978-1-4020-4969-9

- Space-group, science: britannica.com, Retrieved 08 February, 2020
- Prince, E., ed. (2006). International Tables for Crystallography. International Union of Crystallography. doi:10.1107/97809553602060000001. ISBN 978-1-4020-4969-9
- Miller-indices, science: britannica.com, Retrieved 16 June, 2020
- Aroyo, Mois I.; Müller, Ulrich; Wondratschek, Hans (2006). "Historical Introduction". International Tables for Crystallography. A1 (1.1): 2–5. CiteSeerX 10.1.1.471.4170. doi:10.1107/97809553602060000537. Archived from the original on 2013-07-04. Retrieved 2008-04-21

5
Defects in Crystals

In crystallography, crystal defects are the discrepancies in the regular pattern, arrangement and organization of atoms of crystalline solids. Some of the defects are point defects, planar defects, intrinsic and extrinsic defects, Frenkel defects, interstitial defects, vacancy defects, etc. The topics elaborated in this chapter will help in gaining a better perspective about these crystal defects.

A perfect crystal, with every atom of the same type in the correct position, does not exist. All crystals have some defects. Defects contribute to the mechanical properties of metals. In fact, using the term "defect" is sort of a misnomer since these features are commonly intentionally used to manipulate the mechanical properties of a material. Adding alloying elements to a metal is one way of introducing a crystal defect. Nevertheless, the term "defect" will be used, just keep in mind that crystalline defects are not always bad. There are basic classes of crystal defects:

- Point defects, which are places where an atom is missing or irregularly placed in the lattice structure. Point defects include lattice vacancies, self-interstitial atoms, substitution impurity atoms, and interstitial impurity atoms.
- Linear defects, which are groups of atoms in irregular positions. Linear defects are commonly called dislocations.
- Planar defects, which are interfaces between homogeneous regions of the material. Planar defects include grain boundaries, stacking faults and external surfaces.

It is important to note at this point that plastic deformation in a material occurs due to the movement of dislocations (linear defects). Millions of dislocations result for plastic forming operations such as rolling and extruding. It is also important to note that any defect in the regular lattice structure disrupts the motion of dislocation, which makes slip or plastic deformation more difficult. These defects not only include the point and planer defects mentioned above, and also other dislocations. Dislocation movement produces additional dislocations, and when dislocations run into each other it often impedes movement of the dislocations. This drives up the force needed to move the dislocation or, in other words, strengthens the material.

Point Defects

Point defects are where an atom is missing or is in an irregular place in the lattice structure. Point defects include self interstitial atoms, interstitial impurity atoms, substitutional atoms and vacancies. A self interstitial atom is an extra atom that has crowded its way into an interstitial void in the crystal structure. Self interstitial atoms occur only in low concentrations in metals because they distort and highly stress the tightly packed lattice structure.

A substitutional impurity atom is an atom of a different type than the bulk atoms, which has replaced one of the bulk atoms in the lattice. Substitutional impurity atoms are usually close in size (within approximately 15%) to the bulk atom. An example of substitutional impurity atoms is the zinc atoms in brass. In brass, zinc atoms with a radius of 0.133 nm have replaced some of the copper atoms, which have a radius of 0.128 nm.

Interstitial impurity atoms are much smaller than the atoms in the bulk matrix. Interstitial impurity atoms fit into the open space between the bulk atoms of the lattice structure. An example of interstitial impurity atoms is the carbon atoms that are added to iron to make steel. Carbon atoms, with a radius of 0.071 nm, fit nicely in the open spaces between the larger (0.124 nm) iron atoms.

Vacancies are empty spaces where an atom should be, but is missing. They are common, especially at high temperatures when atoms are frequently and randomly change their positions leaving behind empty lattice sites. In most cases diffusion (mass transport by atomic motion) can only occur because of vacancies.

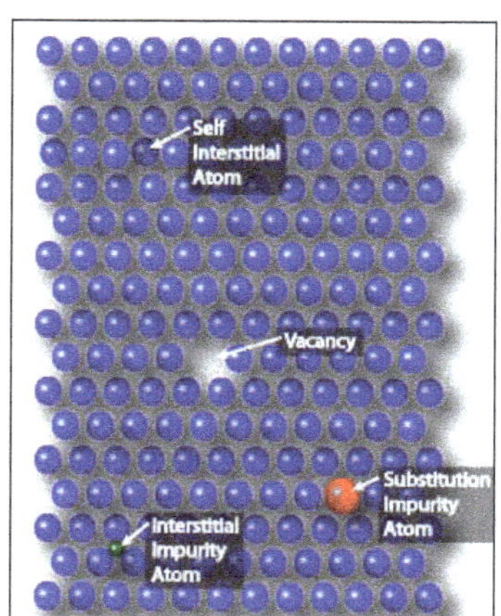

Schottky Defect

A Schottky defect is a type of point defect in a crystal lattice named after Walter H. Schottky. In non-ionic crystals it refers to lattice valency defects.

In ionic crystals, this type of point defect forms when oppositely charged ions leave their lattice sites, creating vacancies. These vacancies are formed in stoichiometric units, to maintain an overall neutral charge in the ionic solid. The surrounding atoms then move to fill these vacancies, causing new vacancies to form. Normally these defects will lead to a decrease in the density of the crystal or metal. Chemical equations in Kröger–Vink notation for the formation of Schottky defects in TiO_2 and $BaTiO_3$.

$$\varnothing \rightleftharpoons v_{Ti}^{''''} + 2v_{O}^{\bullet\bullet}$$

$$\varnothing \rightleftharpoons v_{Ba}^{''} + v_{Ti}^{''''} + 3v_{O}^{''}$$

This can be illustrated schematically with a two-dimensional diagram of a sodium chloride crystal lattice:

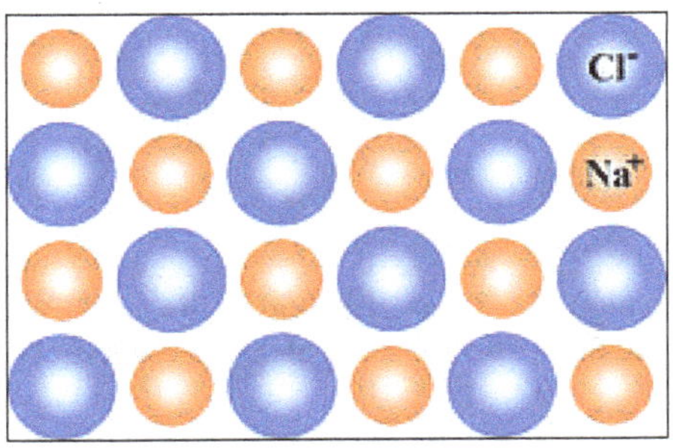

The defect-free NaCl structure.

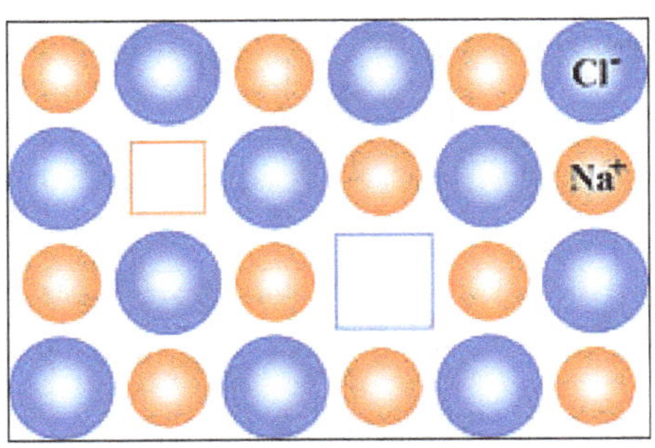

Schottky defects within the NaCl structure.

Schottky defects consist of unoccupied anion and cation sites in a stoichiometric ratio. For a simple ionic crystal of type A^-B^+, a Schottky defect consists of a single anion vacancy (A) and a single cation vacancy (B), or $v_A^{\bullet} + v_B^{'}$ following Kröger–Vink notation. For a more general crystal with formula A_xB_y, a Schottky cluster is formed of x vacancies of A and y vacancies of B, thus the overall stoichiometry and charge neutrality are conserved.

Schottky defects are observed most frequently when there is a small difference in size between the cations and anions that make up a material. They can be produced as the result of the thermal incorporation of unoccupied lattice sites from the exterior of the crystal. The lattice undergoes thermal vibration and thermal expansion when the temperature is raised. When this happens a pair of vacancies is incorporated in the crystal and charge neutrality is maintained. The ion which is missing from the lattice point occupies an interstitial position.

Bound and Dilute Defects

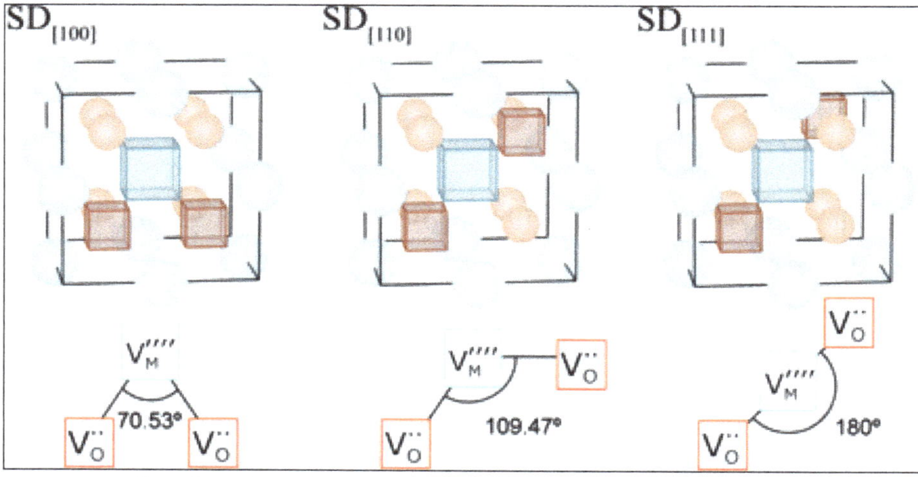

Three bound configurations of Schottky defects in an oxide with Fluorite structure. Spheres represent atoms, cubes represent vacancies.

The vacancies that make up the Schottky defects have opposite charge, thus they experience a mutually attractive Coulomb force. If sufficient thermal energy is available the vacancies may migrate through the crystal lattice, and form bound clusters.

The bound clusters are typically less mobile than the dilute counterparts, as multiple species need to move in a concerted motion for the whole cluster to migrate. This has important implications for numerous functional ceramics used in a wide range of applications, including ion conductors, Solid oxide fuel cells and nuclear fuel.

Examples:

This type of defect is typically observed in highly ionic compounds, highly coordinated compounds, and where there is only a small difference in sizes of cations and anions of which the compound lattice is composed. Typical salts where Schottky disorder is observed are NaCl, KCl, KBr, CsCl and AgBr. For engineering applications, Schottky defects are important in oxides with Fluorite structure, such as CeO_2, cubic ZrO_2, UO_2, ThO_2 and PuO_2.

Effect on Density

Since the total number of ions present in the crystal with this defect is less than the theoretical number of ions for a crystal of its volume, the density of the solid crystal is less than the theoretical density of the material.

Calculation

The number of defects for an MX crystal can be calculated by the given formula:

$$n_s \approx N\,exp\left(-\frac{\Delta H_s}{2RT}\right)$$

where,

- ns = number of Schottky defects per unit volume,
- ΔH = enthalpy of defect formation,
- R = gas constant,
- T = absolute temperature (in K).

We can calculate N by using the formula;

$$N = \frac{density\ of\ the\ ionic\ crystal\ compound \times NA}{molar\ mass\ of\ the\ ionic\ crystal\ compound}$$

Frenkel Defect

A Frenkel defect or dislocation defect is a type of point defect in crystalline solids named after its discoverer Yakov Frenkel. The defect forms when an atom or smaller ion (usually cation) leaves its place in the lattice, creating a vacancy, and becomes an interstitial by lodging in a nearby location. Their primary mechanism of generation is by particle irradiation, as their equilibrium concentration according to the Boltzmann distribution is much smaller than the pure interstitial atoms.

In an ionic crystal, this point defect forms when an ion is displaced from its lattice position to an interstitial site, creating a vacancy at the original site and an interstitial defect at the new location without any changes in chemical properties. This defect appears in an ionic solid which usually possess low coordination number or a considerable disparity in the sizes of the ions. The smaller ion (usually the cation) is dislocated; e.g. in ZnS, AgCl, AgBr, and AgI due to the small size of the zinc ion with charge +2 and the silver ion with charge +1, the cations in the respective cases get dislocated.

AgBr exhibits both Frenkel defect and Schottky defect despite the disparity in the sizes of the constituent ions.

Effect on Density

Frenkel defect does not have any impact on the density of the solid as it involves only the migration of the ions within the crystal, thus preserving both the volume as well as mass.

Examples:

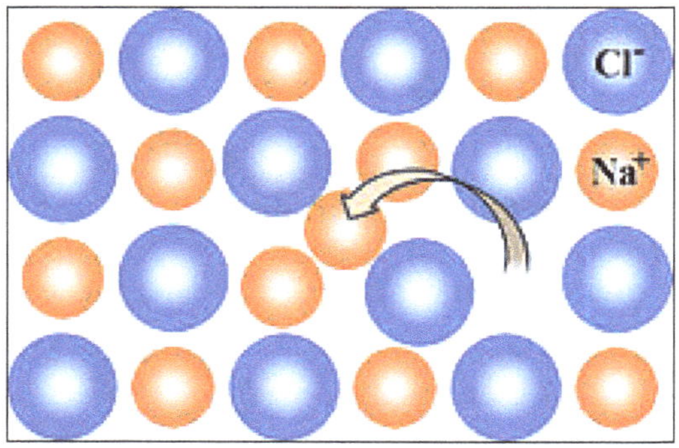

The Frenkel defect within the NaCl structure.

Frenkel defects are exhibited in ionic solids with a large size difference between the anion and cation (with the cation usually smaller due to an increased effective nuclear charge).

Some examples of solids which exhibit Frenkel defects:

- Zinc sulfide,
- Silver(I) chloride,
- Silver(I) bromide (also shows Schottky defects),
- Silver(I) iodide.

These are due to the comparatively smaller size of Zn^{2+} and Ag^{+} ions.

For example, consider a lattice formed by X^{n-} and M^{n+} ions. Suppose an M ion leaves the M sublattice, leaving the X sublattice unchanged. The number of interstitials formed will equal the number of vacancies formed.

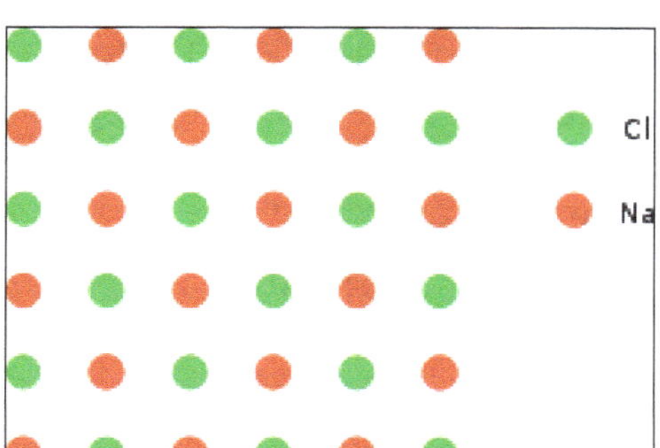

The defect-free NaCl structure.

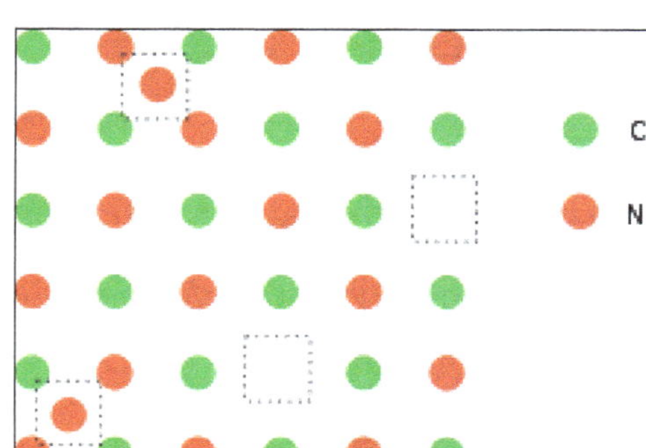

Two Frenkel defects within the NaCl structure.

One form of a Frenkel defect reaction in MgO with the oxide anion leaving the lattice and going into the interstitial site written in Kröger–Vink notation:

$$Mg_{Mg}^{\times} + O_O^{\times} \rightarrow O_i^{''} + v_O^{\bullet\bullet} + Mg_{Mg}^{\times}$$

This can be illustrated with the example of the sodium chloride crystal structure. The diagrams below are schematic two-dimensional representations.

The Frenkel Defect in a Molecule

The Frenkel Defect explains a defect in the molecule where an atom or ion (normally the cation) leaves its own lattice site vacant and instead occupies a normally vacant site. As depicted in the picture below, the cation leaves its own lattice site open and places itself between the area of all the other cations and anions. This defect is only possible if the cations are smaller in size when compared to the anions.

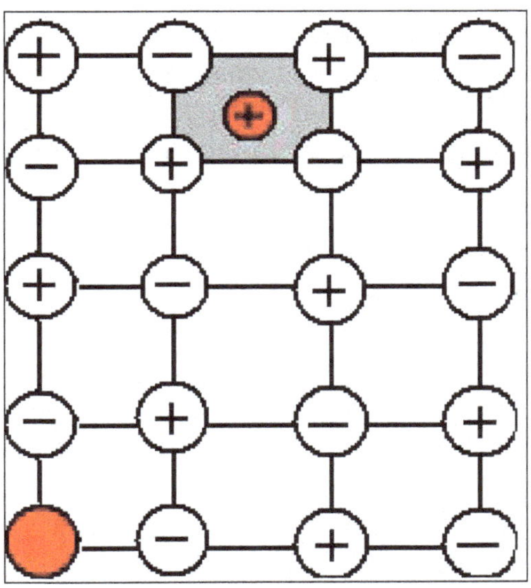

The Frenkel Defect in a molecule.

The number of Frenkel Defects can be calculated using the equation:

$$\sqrt{NN^*}\, e^{\delta \frac{H}{2RT}}$$

where N is the number of normally occupied positions, N^* is the number of available positions for the moving ion, the delta H of formation is the enthalpy formation of one Frenkel defect, and R is the gas constant. Frenkel defects are intrinsic defects because the existence causes the Gibbs energy of a crystal to decrease, which means it's favorable to occur.

Molecules Found with a Frenkel Defect

The crystal lattices are relatively open and the coordination number is low.

Discrete-lattice Theory for Frenkel-defect Aggregation in Irradiated Ionic Solids

It is well known that the primary radiation defects in ionic solids—the F centers ~electron trapped by anion vacancy and the H centers ~interstitial halide atoms—begin to aggregate under intensive irradiation and at sufficiently high temperatures. This leads to the formation of alkali metal colloids and gas bubbles, respectively; a similar process occurs also in heavily irradiated metals. Intensive experimental studies of the conditions for the defect aggregation and subsequent colloid formation ~such as the temperature interval, dose rate, etc. continue nowadays for many alkali halides, in particular for the technologically important CaF_2 , and for ceramic materials. This problem is also interesting from a fundamental point of view, being an example of pattern formation and self-organization in reaction-diffusion systems far from equilibrium. It should be stressed that metal colloid formation is a rather slow process which in real experiments often lasts several weeks or longer; another reason why this problem is a difficult one for direct computer ~Monte Carlo simulations is that the mobilities of the two kinds of defects involved ~interstitials and vacanciesdiffer typically by 10–15 orders of magnitude. In order to model the F center aggregation, one has to use a very small time increment which is dictated by the highly mobile H centers.

Existing theories of radiation-induced defect aggregation and colloid formation can be classified, in terms of the mathematical formalism used, into three categories: macroscopic, mesoscopic, and microscopic. We call microscopic a theory on an atomic scale which uses no fitting or uncertain parameters like numerous reaction rates but only several basic defect parameters such as the diffusion energies and interaction energies. Such a theory has been presented recently by us. However, the continuum approximation used there has led to nonphysical problems associated with cutting off the interaction potentials at the nearestneighbor ~NN distance, and with large gradients of the correlation functions at these distances. The first point makes results very sensitive to the manner in which the potentials are defined at distances shorter than NN. The second one forces us to use very small coordinate increments in the calculations which then strongly restricts the time interval which can realistically be reached using such an approach.

Physical Model

The physical model includes creation of interstitials, i.e., H centers, and vacancies, i.e., F centers called hereafter just defects or particles A and B). This takes place with a given doserate p, AB pairs are not spatially correlated at birth and recombine when they approach each other during their migration to within the NN distance. Therefore, their macroscopic concentrations always coincide, $n = n_A = n_B$. Isolated singledefects hop with the activation energy El and are characterized by the diffusion coefficients $D_\lambda = d_\lambda \exp(-E_\lambda / k_B T), \lambda = A, B$. When several defects are close together, the hop rate of a given defect to the nearest empty lattice site is determined by both the local defect configuration and the interaction between defects; this can change its effective diffusion coefficient D_{λ}^{eff} by many orders of magnitude compared to that for a single defect. This affects the effective reaction rate K of the A and B recombination; for the dilute system in the continuum approximation this rate is well known to be $K_0 = 4\pi r_0 (D_A + D_B)$, where r_o is the recombination radius.

From previous theoretical studies it follows that defect attraction plays a decisive role in the aggregation process. It is incorporated in our model via three types of NN attractions between the two types of NN defects in the spirit of the Ising model: E_{AA}, E_{AB}, and E_{BB}. Note that only a few theoretical studies have been devoted up to now to the effects of particle interaction especially, for similar particles in the kinetics of the bimolecular diffusion-controlled reaction $A + B \rightarrow 0$, with the emphasis on fluctuation phenomena. Particle interactions render the kinetic equations essentially nonlinear and this causes manifestation of self-organization pattern formation under irradiation.

Mathematical Model

Each lattice site is given a lattice vector r and coordination number z. The state of the site r is described by the local variable σ_l. The spectrum of its values $\sigma_l = 0, A, B$ describes three possible states of the lattice site: o represents a vacant site, whereas A and B are sites occupied by an A or B particle, respectively. A particle aggregate or colloid is treated as a dynamical domain containing particles of one type only (A or B). In other words, an isolated particle A or particle A inside the aggregate are described in a similar way through the local state variable .

State Probabilities

The lattice processes under consideration can be treated as stochastic Markov processes. Each lattice state characterized by a set of local variables $\sigma = \{\sigma_1, \sigma_2, ...\}$ arises with the probability $\rho(\sigma)$. The latter gives formally a complete probabilistic description of the system, however the equation determining $\rho(\sigma)$ cannot be solved exactly. The probabilistic description of the lattice states can be done also using another set of quantities, the so-called many-point probabilities which depend on i lattice sites and are coupled through recurrence sum rules. Such a limited set of probabilities better suits our particular problem.

Single-point Probabilities

The single-point probability $\rho^{(1)}(\sigma_l)$ is the probability to find at time t the lattice site r_1 in the state σ_l. Due to the translational invariance these probabilities can depend only on time. Using the abbreviation $\lambda = \sigma_l, C_\lambda = \rho^{(1)}(\sigma_l)$ one can write down the obvious normalization condition:

$$\sum_{\lambda} C_\lambda = C_A + C_B + C_0 = 1$$

In terms of physics the quantity C_λ is nothing but a simple density of particles on the lattice site population. C_A, C_B, and C_o are densities for particles A,B or empty sites, respectively. The usual, dimensional densities n_λ per unit volume are just:

$$n_\lambda = \frac{C_\lambda}{a_0^3}$$

where a_o is a lattice parameter. Obviously, the use of such macroscopic values as the simple particle densities C_λ is not sufficient for the description of the particle aggregation. The latter needs a detailed analysis of the relative distribution of several types of particles on the lattice and an analysis of their relevant spatial correlations. For this purpose the twopoint densities and relevant correlation functions are well suited.

The Correlation Functions

The two-point probability $\rho^{(2)}(\sigma_l, \sigma_m)$ is the probability of finding at the time t the site r_l in the state $\lambda = \sigma_l$ and the site $\boldsymbol{r}_m$ in the state $\mu = \sigma_m$. It can be formally presented in the factorized form:

$$\rho^{(2)}(\sigma_l, \sigma_m) = C_\lambda C_\mu F_{\lambda\mu}(\boldsymbol{r}), \quad \boldsymbol{r} = \boldsymbol{r}_l - \boldsymbol{r}_m$$

Where $F_{\lambda\mu}(\boldsymbol{r})$ is the joint correlation function. As follows from $\rho^{(2)}(\sigma_l, \sigma_m) = C_\lambda C_\mu F_{\lambda\mu}(\boldsymbol{r})$, $\boldsymbol{r} = \boldsymbol{r}_l - \boldsymbol{r}_m$ due to the translational invariance, both the two-point density and the correlation function depend only on the relative distance between two sites $\boldsymbol{r}_l$ and $\boldsymbol{r}_m$, characterized by a vector **r**. Besides, $F_{\lambda\mu}(\boldsymbol{r})$ satisfies several general conditions. The first one is that the correlation disappears at large distances ~due to the short-range nature of the spatial correlations!. This means that as $r \to \infty$ the states of sites $\boldsymbol{r}_l$ and $\boldsymbol{r}_m$ are statistically independent which results in:

$$\rho^{(2)}(\sigma_l, \sigma_m) = \rho^{(1)}(\sigma_l)\rho^{(1)}(\sigma_m) \text{ , i.e.,}$$

$$\lim F_{\lambda\mu}(\boldsymbol{r}) = 1$$

The second condition is a relation between single- and twopoint densities, the sum rule:

$$\sum_{\sigma_m} \rho^{(2)}(\sigma_l, \sigma_m) = \rho^{(1)}(\sigma_l),$$

Or,

$$\sum_{\mu} c_\mu^{(\lambda)}(\boldsymbol{r}) = 1, \; c_\mu^{(\lambda)}(\boldsymbol{r}) = C_\mu F_{\lambda\mu}(\boldsymbol{r}).$$

The latter equation is similar to $\sum_{\lambda} C_{\lambda} = C_A + C_B + C_0 = 1$ and permits also a similar interpretation. The quantity:

$$c_{\mu}^{(\lambda)}(\boldsymbol{r}) = \rho^{(2)}(\sigma_l, \sigma_m) / \rho^{(1)}(\sigma_l),$$

defines the conditional probability that the site at r is in the state μ when the central site r_1 is in the state λ.

Density of Single Particles

To handle the complicated statistical problem under question, we use a reduced probabilistic description of the many particle system in terms of single and joint particle densities only, i.e., many-point probabilities, $\rho^{(k)}$ are expressed through simple densities concentrations C_{λ} and the joint correlation functions $F_{\lambda\mu}(\boldsymbol{r})$. Let us illustrate this principle by an example. We call an isolated particle single center a local state where the site $\mathbf{r}_1$ is in the state $\lambda = \sigma_l$ whereas all its nearest-neighbor NN sites are empty $\left(\{\sigma\}_l^z = \{\sigma_1, ..., \sigma_z\} = \{0, ..., 0\}\right)$. The probability to find an isolated single particle C_{λ}^s is defined by the $(z+1)$-point probability $C_{\lambda}^s = \rho^{(z+1)}$. The density of dimer-, trimer- and higher particle configurations can be found in a similar way, i.e., through many-particle densities with a larger number of arguments ~three, four etc. As a result, one arrives at an infinite set of coupled equations for the correlation functions of all orders which cannot be solved exactly. An approximation to handle this problem is to cut off this infinite hierarchy. Single densities can be calculated using the cluster approximation well known in theory of phase transitions. The approximation is based on the relation:

$$\rho^{(z+1)} = C_{\lambda} \prod_{i=1}^{z} c_{\mu_i}^{(\lambda)}(1).$$

In other words, the probability of finding a cluster of $(z+1)$ sites is approximated as a product of the probability C_{λ} to find the central site in the λ state and the conditional probabilities to find NN sites in the states $\mu_i = \sigma_i$ (its argument $r = 1$). From this general expression for the isolated particles one gets:

$$C_{\lambda}^s = C_{\lambda}\left(c_0^{(\lambda)}(1)\right)^z$$

Since the conditional probabilities are limited from above equation $\sum_{\mu} c_{\mu}^{(\lambda)}(\boldsymbol{r}) = 1,\ c_{\mu}^{(\lambda)}(\boldsymbol{r}) = C_u F_{\lambda\mu}(\boldsymbol{r})$, as expected the relation $C_{\lambda}^s \le C_{\lambda}$ (holds not all particles are isolated). It will be shown below that by incorporation of particle attraction $(E_{\lambda\lambda} < 0)$ this non-equality can be strengthened, $C_{\lambda}^s << C_{\lambda}$ due to formation of dimers, trimers, and higher configurations. Similar to equation $n_{\lambda} = \frac{C_{\lambda}}{a_0^3}$ the dimensional density of isolated particles can be found as,

$$n_{\lambda}^{(1)} = \frac{C_{\lambda}^{s}}{a_0^3}$$

The Density of Dimer Particles

In the dimer configuration, the two NN particles are characterized by the lattice vectors $\mathbf{r}_l$ and $\mathbf{r}_m$ and the states, $\lambda = \upsilon = A, B$ whereas all surrounding sites are empty. Such a state is described by the (2z)-point probability $\rho^{(2z)}$ provided that the states of the central sites are $\sigma_l = \sigma_n = \lambda$ whereas those of $(z-1)$ NN sites surrounding each of the two central sites are $\{\sigma\}_n^{z-1} = \{\sigma_1, ..., \sigma_{z-1}\} = \{0, ..., 0\}$, $\{\sigma\}_n^{z-1} = \{\sigma_1^{'}, ..., \sigma_{z-1}^{'}\} = \{0, ..., 0\}$ respectively. Note that this is true for cubic lattices only. However, our formalism can be generalized for any lattice structures. Using the cluster approximation, equation $\rho^{(z+1)} = C_{\lambda} \prod_{i=1}^{z} c_{\mu_i}^{(\lambda)}(1)$. We arrive at,

$$\rho^{(2z)} = C_{\lambda} C_{\upsilon} F_{\lambda\upsilon}(1) \prod_{i=1}^{z-1} c_{\mu_i}^{(\lambda)}(1) \prod_{j=1}^{z-1} c_{\mu_j}^{(\upsilon)}(1)$$

Taking into account equaition. $\rho^{(2z)} = C_{\lambda} C_{\upsilon} F_{\lambda\upsilon}(1) \prod_{i=1}^{z-1} c_{\mu_i}^{(\lambda)}(1) \prod_{j=1}^{z-1} c_{\mu_j}^{(\upsilon)}(1)$ and all possible spatial orientations of the dimer a dumbbell, one gets:

$$C_{\lambda}^{d} = \frac{z}{2} C_{\lambda}^{2} F_{\lambda\lambda}(1) \left(c_0^{(\lambda)}(1)\right)^{2(z-1)}$$

The density of dimers is defined in such a way that the dimensional concentration $n_{\lambda}^{(2)} = C_{\lambda}^{d} / a_0^3$ gives a mean number of dimers per unit volume.

The Mean Particle Number and Aggregate Size

In a similar manner the density of trimers, tetramers, and larger aggregates can be found. However, large aggregates colloids can be better characterized by such averaged quantities as aggregate size R_{λ} and the mean number of particles inside it, N_{λ} For this purpose we can use the following arguments. By definition, $c_{\lambda}^{(\lambda)}(\boldsymbol{r})$ is the mean density of particles λ at the distance r from the central site occupied by a similar particle λ. If the particle distribution is random, $F_{\lambda\mu}(\boldsymbol{r}) = 1$, the local density coincides with the macroscopic one, $c_{\lambda}^{(\lambda)}(\boldsymbol{r}) = C_{\lambda}$. However, this is only the case if there are no particle interactions or reaction. In the general case the relations, $F_{\lambda\mu}(\boldsymbol{r})\to 1$ and $c_{\lambda}^{(\lambda)}(\boldsymbol{r})=C_{\lambda}$ hold only asymptotically, as $r\to\infty$. This is why the quantity $\delta c_{\lambda}^{(\lambda)}(\boldsymbol{r})=C_{\lambda}[F_{\lambda\lambda}(\boldsymbol{r})\to 1]$ gives the surplus particle density at short relative distances as compared to their random distribution. Its positive magnitude characterizes the aggregation of particles. Using $\delta c_{\lambda}^{(\lambda)}(\boldsymbol{r})$ as a weighting function, we can define the mean number of particles inside the aggregate as:

$$N_{\lambda}=1+\sum_{\boldsymbol{r}}\delta c_{\lambda}^{(\lambda)}(\boldsymbol{r})$$

and its radius,

$$R_{\lambda}=\frac{1}{N_{\lambda}}\sum_{\boldsymbol{r}}|\boldsymbol{r}|\delta c_{\lambda}^{(\lambda)}(\boldsymbol{r})$$

The unity on the right-hand side of $N_{\lambda}=1+\sum_{\boldsymbol{r}}\delta c_{\lambda}^{(\lambda)}(\boldsymbol{r})$ corresponds to the central particle in whose surroundings the excess particle density is integrated. For a random particle distribution, $F_{\lambda\mu}(\boldsymbol{r})=1$ and the definitions $N_{\lambda}=1+\sum_{\boldsymbol{r}}\delta c_{\lambda}^{(\lambda)}(\boldsymbol{r})$ and $R_{\lambda}=\frac{1}{N_{\lambda}}\sum_{\boldsymbol{r}}|\boldsymbol{r}|\delta c_{\lambda}^{(\lambda)}(\boldsymbol{r})$ yield $N_{\lambda}=1$ and $R_{\lambda}=0$, i.e., the concentration of dimers, trimers, etc. which always statistically exist, is assumed to be much smaller than the aggregation effect driven by the reaction and particle interaction. It should be noted here that in real experiments in insulating crystals concentrations of isolated centers, dimers, and trimers the so-called F, F_2 (M), F_3 (R) centers, i.e., the smallest aggregates, on the one hand, and the large colloid size and a total concentration of metal fraction of sites occupied by defects, on the other hand, can be estimated from absorption measurements. Information about H aggregates is much less accurate, only the smallest H clusters are detectable by Raman spectroscopy.

Transition Rates

After definition of the spectrum of the site state variables $\sigma_l=\lambda$ the stochastic lattice Markovian model of the aggregation kinetics needs also the definition of the transition rates. The latter are monomolecular and bimolecular transitions.

Monomolecular Transitions

The simplest, monomolecular transitions depend only on a single lattice site state whereas the states of surrounding sites remain unchanged. These transitions can be presented as:

$$\sigma_l\xrightarrow{k}\sigma_l' \; or \; \lambda\xrightarrow{k}\lambda'$$

where the transition rate $k = P(\lambda \xrightarrow{k} \lambda')$ does not depend on the lattice vector r_1. Examples are A (B) particle defect creation in an empty lattice site, $0 \to A$ *and* $0 \to B$, respectively:

$$P(0 \to A) = P(0 \to B) = p_0 = pa_0^3$$

where p is defined in Sec. III as the irradiation intensity dose rate. That is, Frenkel defects are created in pairs, each occupying two lattice sites. Depending on the irradiation conditions, the distance within such a pair can vary. Under an electron irradiation considered here it is known to be quite large, so that dissimilar defect creation can be considered as a random, uncorrelated creation of A and B particles in equal concentrations. In principle, the spatial correlation of Frenkel defect pairs, AB, can also be taken into account but it is not directly related to the problem of aggregation kinetics.

Bimolecular Transitions

Such transitions are related to a simultaneous change of the state of two NN sites with lattice vectors $\mathbf{r}_l$ and $\mathbf{r}_n$. In the general case bimolecular transitions are described by the equation,

$$\sigma_l \sigma_n \xrightarrow{k} \sigma_l' \sigma_n' \quad or \quad \lambda v \xrightarrow{k} \lambda' v'$$

It is convenient to present the transition rate k taking into account the coordination number z: $k = K(\lambda v \to \lambda' v')/z$.

Examples are: Dissimilar particle recombination,

$$AB \xrightarrow{k} 00 \quad with \quad K(AB \to 00) = \Omega$$

particle diffusion (particle random walks on the lattice, $\lambda = A, B$),

$$\lambda 0 \xrightarrow{k} 0\lambda \quad with \quad K(\lambda 0 \to 0\lambda) = D_\lambda^0$$

For simplicity we assume the recombination rate to be infinite, $\Omega \to \infty$, which corresponds to the instant recombination of defects with a restoration of the perfect lattice. The hop rates D_A^0 *and* D_B^0 have simple relations to the diffusion coefficients.

$$D_\lambda = \frac{a_0^2}{z} D_\lambda^0, \quad D_\lambda = \exp\left(-\frac{E_\lambda}{k_B T}\right),$$

where d_λ is the diffusion prefactor and a_0 is the hopping distance assumed to be equal to the lattice constant.

Energetic Interaction of Particles

Equations $\sigma_l \sigma_n \xrightarrow{k} \sigma_l' \sigma_n'$ *or* $\lambda v \xrightarrow{k} \lambda' v'$, $AB \xrightarrow{k} 00$ *with* $K(AB \to 00) = \Omega$, and

$\lambda 0 \xrightarrow{k} 0\lambda$ *with* $K(\lambda 0 \to 0\lambda) = D_\lambda^0$ describe particle creation, diffusion, and recombination but neglect their energetic interaction. Due to this interaction the transition rate can depend on the NN states and temperature. Examples of such interactions are Coulomb and elastic attraction of defects in solids. To take this into account, the monomolecular transition rate $P(\lambda \to \lambda')$ should be replaced by $\boldsymbol{P}\left(\lambda \to \lambda' \middle| \{\sigma\}_l^z\right)$ and the bimolecular rate $K(\lambda\upsilon \to \lambda'\upsilon')$ by $\boldsymbol{K}\left(\lambda\upsilon \to \lambda'\upsilon' \middle| \{\sigma\}_l^{z-1}\{\sigma\}_n^{z-1}\right)$. This dependence can be found from the asymptotic behavior of the steady state of the kinetic system under study and the equilibrium Gibbs distribution. This problem arose in the study of the kinetic Ising model in lattice phase transitions. We follow here the general approach developed by us for the stochastic surface reactions. The so-called standard model presented there allows us to obtain uniquely the rates of mono- and bimolecular transitions including the particle interactions. It should be stressed once more that we restrict ourselves to NN interaction. Due to the interaction similar particles begin to aggregate. It is clear that the probability of some particle in the center of an aggregate to find an empty NN site for a diffusion hop is quite small. Also, the hop probability for a particle at the periphery of the aggregate to leave it is because of the particle attraction smaller than the probability to hop into the interior of the aggregate and make it denser. This is why the aggregation kinetics should depend on temperature and the three interaction energies between similar and dissimilar particles, E_{AA}, E_{AB}, E_{BB}.

A system of interacting particles can be characterized by the effective diffusion coefficient D_λ^{eff}. By definition, equation. $\lambda 0 \xrightarrow{k} 0\lambda$ *with* $K(\lambda 0 \to 0\lambda) = D_\lambda^0$, rates of bimolecular hops are proportional to the diffusion coefficients. Let us define the effective dimensional diffusion coefficient, $D_\lambda^{eff} / D_\lambda$. Here D_λ is the usual diffusion coefficient in a dilute system consisting of isolated defects which is proportional to the rate $K(\lambda 0 \to 0\lambda)$. In turn, in the dense system of interacting particles D_λ^{eff} is proportional to the averaged transition rate $\left\langle \boldsymbol{K}\left(\lambda 0 \to 0\lambda \middle| \{\sigma\}_l^{z-1}\{\sigma\}_n^{z-1}\right)\right\rangle$. The latter transition rate is averaged $\langle(....)\rangle$ over all configurations of particles surrounding a given particle. This can be done using the cluster probability, equation $\rho^{(2z)} = C_\lambda C_\upsilon F_{\lambda\upsilon}(1)\prod_{i=1}^{z-1} c_{\mu_i}^{(\lambda)}(1)\prod_{j=1}^{z-1} c_{\mu_j}^{(\upsilon)}(1)$. Under particle aggregation the parameter $D_\lambda^{eff} / D_\lambda$ is by many orders of magnitude less than unity thus characterizing migration properties in a dense interacting system.

Master Equations

The stochastic Markovian process is uniquely defined by a set of mono- and bimolecular transition rates. The relevant, discretelattice formalism based on the hierarchy of the correlation functions has been presented recently and illustrated by the particle aggregation in the course of the catalytic surface reaction $A+\frac{1}{2}B_2 \to 0$. In contrast, we discuss only the basic idea of the approach and its application to the particular bulk $A+B \to 0$ reaction in three dimensions. In fact, the latter reaction under study is mathematically isomorphic to the abovementioned surface catalytic reaction where the molecule adsorption/desorption is replaced by the defect production rate p in the bulk. The final set of kinetic equations read:

$$\frac{dC_\lambda}{dt} = G_\lambda(C,F),$$

$$\frac{dF_{\lambda\mu}(\boldsymbol{r})}{dt} = G_{\lambda\mu}(C,F),$$

Where G_λ and $G_{\lambda\mu}$ are expressions nonlinear in the lattice densities and the correlation functions. They arise due to the systematic use of approximations such as the cluster approximation, equation $\rho^{(2z)} = C_\lambda C_v F_{\lambda v}(1)\prod_{i=1}^{z-1} c_{\mu_i}^{(\lambda)}(1)\prod_{j=1}^{z-1} c_{\mu_j}^{(v)}(1)$. To illustrate this approach, we write down the structure of equation $\frac{dC_\lambda}{dt} = G_\lambda(C,F)$, where:

$$G_\lambda = G_\lambda^{in} - G_\lambda^{out},$$

$$G_\lambda^{in} = \sum_{\lambda'} P'(\lambda' \to \lambda)C_{\lambda'} + \sum_{\lambda'v'v} K'(\lambda'v' \to \lambda v)F_{\lambda'v'}(1)C_{\lambda'}C_{v'}$$

$$G_\lambda^{out} = \sum_{\lambda'} P'(\lambda' \to \lambda)C_\lambda + \sum_{\lambda'v'v} K'(\lambda v \to \lambda'v')F_{\lambda v}(1)C_\lambda C_v$$

The equations above describe all possible mono- and bimolecular processes. Formally these equations remind us of the usual rate equations for defect concentrations. However, the key factor is that the functions $P'(\lambda' \to \lambda)$ and $K'(\lambda v \to \lambda'v')$ are effective transition rates. These quantities are obtained via averaging over the nearest-neighbor states based on the statistical weights: Equation $\rho^{(z+1)} = C_\lambda \prod_{i=1}^{z} c_{\mu_i}^{(\lambda)}(1)$ for the calculation of quantity $P'(\lambda' \to \lambda) = \left\langle \boldsymbol{P}'\left(\lambda' \to \lambda \middle| \{\sigma\}_l^z\right)\right\rangle$ and equation $\rho^{(2z)} = C_\lambda C_v F_{\lambda v}(1)\prod_{i=1}^{z-1} c_{\mu_i}^{(\lambda)}(1)\prod_{j=1}^{z-1} c_{\mu_j}^{(v)}(1)$ for the quantity $K'(\lambda v \to \lambda'v') = \left\langle \boldsymbol{K}'\left(\lambda v \to \lambda'v' \middle| \{\sigma\}_l^z - \{\sigma\}_l^{z-1}\right)\right\rangle$ below:

$$P'(\lambda' \to \lambda) = \left\langle \boldsymbol{P}'\left(\lambda' \to \lambda \middle| \{\sigma\}_l^z\right)\right\rangle$$

$$K'(\lambda v \to \lambda' v') = \left\langle K'\left(\lambda v \to \lambda' v' \middle| \{\sigma\}_l^z - \{\sigma\}_l^{z-1}\right)\right\rangle$$

As a result, the kinetic equations depend on the kinetic constants, temperature and NN interaction energies. The particle aggregation is characterized by the shape of the joint correlation functions; at distances $\boldsymbol{r} = |\boldsymbol{r}_l - \boldsymbol{r}_m|$, where some joint density considerably exceeds the asymptotic value of unity, a large excess of neighboring similar particles i.e., an aggregate! is found compared to the random Poisson distribution. The aggregate size can be estimated as the distance at which the joint correlation functions $F_{\lambda\lambda}(\boldsymbol{r})$ approach the value of unity. Our stochastic formalism operates with a finite set of lattice densities C_λ and joint correlation functions $F_{\lambda\mu}(\boldsymbol{r})$.

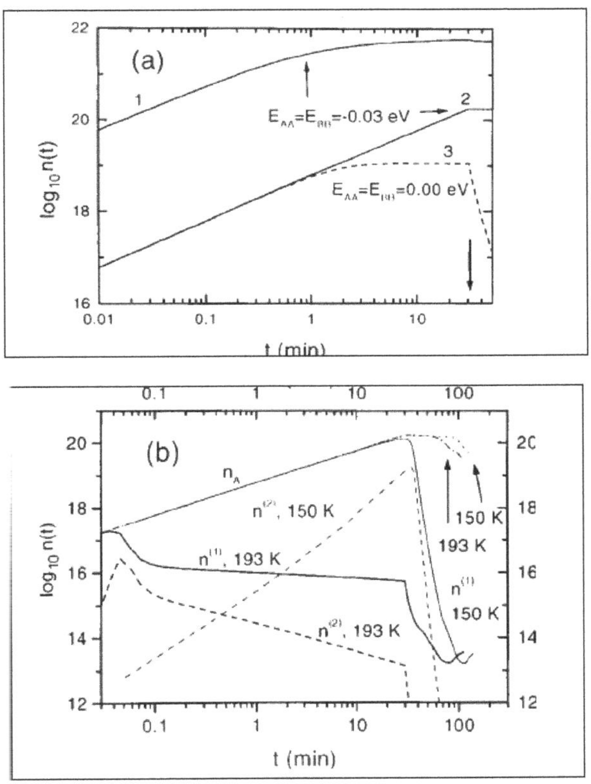

Time development of total F, H concentrations in cm^{-3}) under irradiation at 193 K for 30 min shown by an arrow with a subsequent heating at the rate 1.3 K/min. Curves 1 and 2 show the effect of the dose rate ($p = 10^{20}$ *and* $p = 10^{17}\, cm^{-3}s^{-1}$ respectively), whereas curve 3 shows the effect of the neglect in the latter case of similar defect interation $E_{\lambda\lambda} = 0$, $\lambda = A, B$. Note rapid defect recombination in case 3 when irradiation is switched off. The dynamics of single, $n^{(1)}$, *and* dimer, $n^{(2)}$, H center aggregation and total H concentrations, n_A, for a dose rate $p = 10^{17}\, cm^{-3}s^{-1}$ under irradiation at 193 K as in curve 2 above and at 150 K.

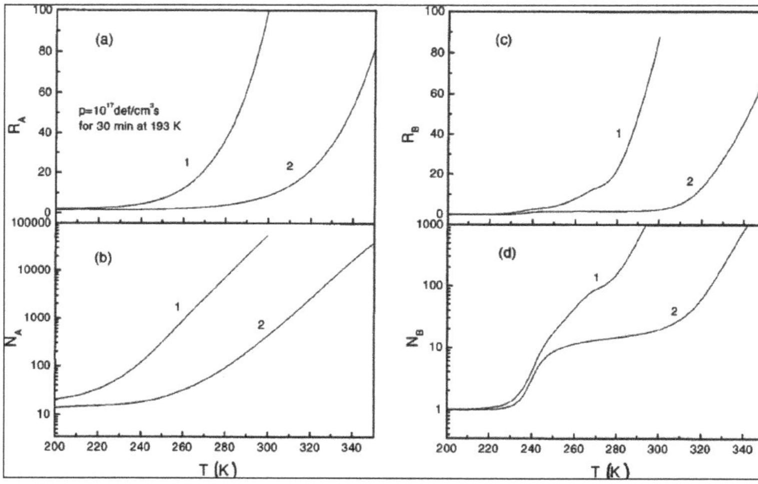

Heating-induced growth of the mean radius of H aggregates, R_A (in units of the $F-F$ distance, $a_o = 2.73\overset{\circ}{A}$ in CaF_2). Note that H and F defects are denoted as defects A and B, respectively. b) The mean number of particles inside each aggregate, N_A. Irradiation with the dose rate $p = 10^{17}\,cm^{-3}s^{-1}$ for 30 min. Curves 1 and 2 correspond to the attraction energies $E_{AA} = E_{BB} = -0.02eV$. and 20.03 eV, respectively; c and d the same for the F centers, respectively.

The set of resulting ordinary differential equations to be calculated is formally infinite due to the infinite number of lattice sites $\mathbf{r}_l$ and $\mathbf{r}_m$; but in practice we restrict ourselves to finite relative distances correlation lengths which is justified by the short-range nature of the spatial correlations in the particle densities.

Interstitial Defect

Interstitials defects are a variety of crystallographic defects where atoms assume a normally unoccupied site in the crystal structure. In interstitial defects, two or more atoms may share one lattice site, thereby increasing its total energy. Alternatively small atoms in some crystals may occupy interstitial sites in energetically favorable configurations, such as hydrogen in palladium. Interstitials can be produced by bombarding a crystal with elementary particles having energy above the displacement threshold for that crystal, but they may also exist in small concentrations in thermodynamic equilibrium.

Self-interstitials

Self-interstitial defects are interstitial defects which contain only atoms which are the same as those already present in the lattice. The structure of interstitial defects has been experimentally determined in some metals and semiconductors.

Contrary to what one might intuitively expect, most self-interstitials in metals with a known structure have a 'split' structure, in which two atoms share the same lattice site. Typically the center of mass of the two atoms is at the lattice site, and they are displaced symmetrically from it along one of the principal lattice directions. For instance, in several common face-centered cubic (fcc) metals such as copper, nickel and platinum, the ground state structure of the self-interstitial is the split [100] interstitial structure, where two atoms are displaced in a positive and negative [100] direction from the lattice site. In body-centered cubic (bcc) iron the ground state interstitial structure is similarly a [110] split interstitial.

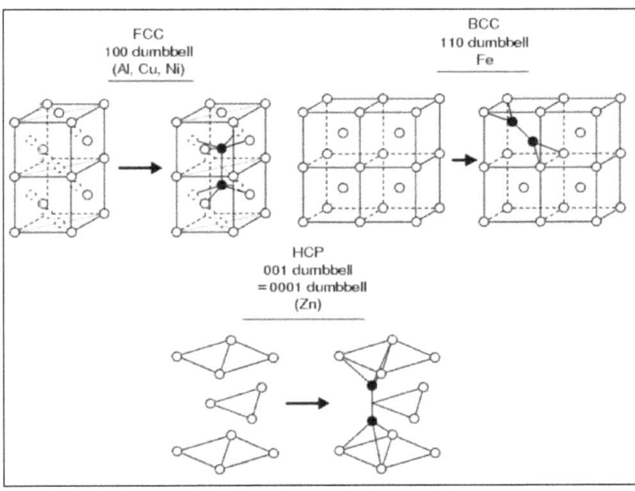

Structure of self-interstitial in some common metals. The left-hand side of each crystal type shows the perfect crystal and the right-hand side the one with a defect.

These split interstitials are often called dumbbell interstitials, because plotting the two atoms forming the interstitial with two large spheres and a thick line joining them makes the structure resemble a dumbbell weight-lifting device.

In other bcc metals than iron, the ground state structure is believed based on recent density-functional theory calculations to be the [111] crowdion interstitial, which can be understood as a long chain (typically some 10–20) of atoms along the [111] lattice direction, compressed compared to the perfect lattice such that the chain contains one extra atom.

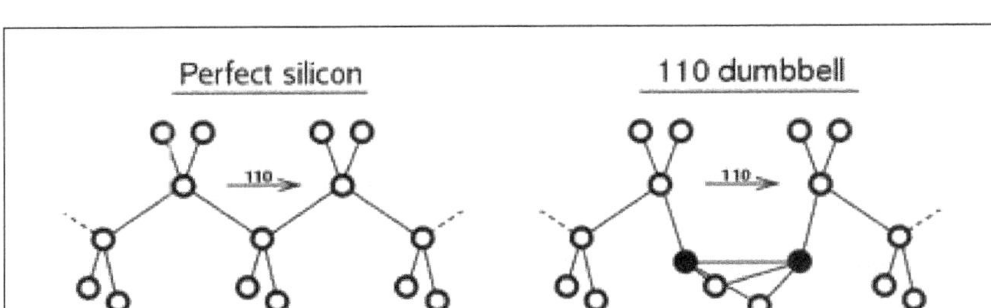

Structure of dumbbell self-interstitial in silicon. The structure of the interstitial in silicon may depend on charge state and doping level of the material.

In semiconductors the situation is more complex, since defects may be charged and different charge states may have different structures. For instance, in silicon, the interstitial may either have a split [110] structure or a tetrahedral truly interstitial one.

Carbon, notably in graphite and diamond, has a number of interesting self-interstitials - recently discovered using LDF-calculations is the "spiro-interestitial" in graphite, named after spiropentane, as the interstitial carbon atom is situated between two basal planes and bonded in a geometry similar to spiropentane.

Impurity Interstitials

Small impurity interstitial atoms are usually on true off-lattice sites between the lattice atoms. Such sites can be characterized by the symmetry of the interstitial atom position with respect to its nearest lattice atoms. For instance, an impurity atom I with 4 nearest lattice atom A neighbours (at equal distances) in an fcc lattice is in a tetrahedral symmetry position, and thus can be called a tetrahedral interstitial.

Large impurity interstitials can also be in split interstitial configurations together with a lattice atom, similar to those of the self-interstitial atom.

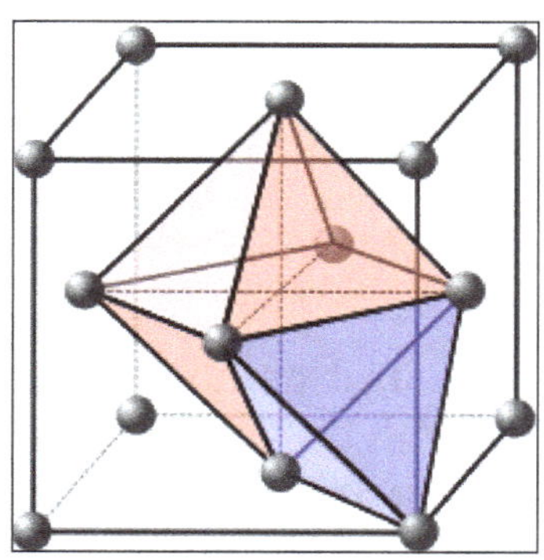

Octahedral (red) and tetrahedral (blue) interstitial symmetry polyhedra in a face-centered cubic lattice. The actual interstitial atom would ideally be in the middle of one of the polyhedra.

Effects of Interstitials

Interstitials modify the physical and chemical properties of materials.

- Interstitial carbon atoms have a crucial role for the properties and processing of steels, in particular carbon steels.
- Impurity interstitials can be used e.g. for storage of hydrogen in metals.
- The crystall lattice can expand with the concentration of impurity interstitials.
- The amorphization of semiconductors such as silicon during ion irradiation is often explained by the buildup of a high concentration of interstitials leading eventually to the collapse of the lattice as it becomes unstable.

- Creation of large amounts of interstitials in a solid can lead to a significant energy buildup, which on release can even lead to severe accidents in certain old types of nuclear reactors (Wigner effect). The high-energy states can be released by annealing.
- At least in fcc lattice, interstitials have a large diaelastic softening effect on the material.
- It has been proposed that interstitials are related to the onset of melting and the glass transition.

Vacancy Defect

In crystallography, a vacancy is a type of point defect in a crystal. Crystals inherently possess imperfections, sometimes referred to as crystalline defects. A defect in which an atom is missing from one of the lattice sites is known as a "vacancy" defect.

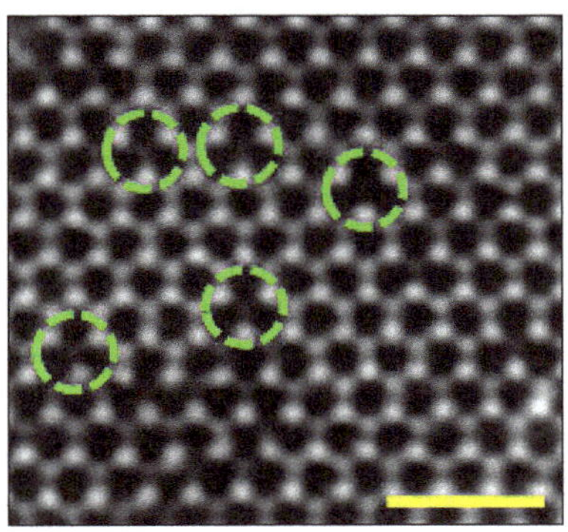

Electron microscopy of sulfur vacancies in a monolayer of molybdenum disulfide. Right circle points to a divacancy, i.e., sulfur atoms are missing both above and below the Mo layer. Other circles are single vacancies, i.e., sulfur atoms are missing only above or below the Mo layer. Scale bar: 1 nm.

Vacancies occur naturally in all crystalline materials. At any given temperature, up to the melting point of the material, there is an equilibrium concentration (ratio of vacant lattice sites to those containing atoms). At the melting point of some metals the ratio can be approximately 1:1000. This temperature dependence can be modelled by:

$$N_v = N\exp(-Q_v / k_B T)$$

where N_v is the vacancy concentration, Q_v is the energy required for vacancy formation, k_B is the Boltzmann constant, T is the absolute temperature, and N is the concentration of atomic sites i.e.,

$$N = \rho N_A / A$$

where ρ is density, N_A Avogadro constant, and A the atomic mass.

It is the simplest point defect. In this system, an atom is missing from its regular atomic site. Vacancies are formed during solidification due to vibration of atoms, local rearrangement of atoms, plastic deformation and ionic bombardments.

The creation of a vacancy can be simply modeled by considering the energy required to break the bonds between an atom inside the crystal and its nearest neighbor atoms. Once that atom is removed from the lattice site, it is put back on the surface of the crystal and some energy is retrieved because new bonds are established with other atoms on the surface. However, there is a net input of energy because there are fewer bonds between surface atoms than between atoms in the interior of the crystal.

Stacking Fault

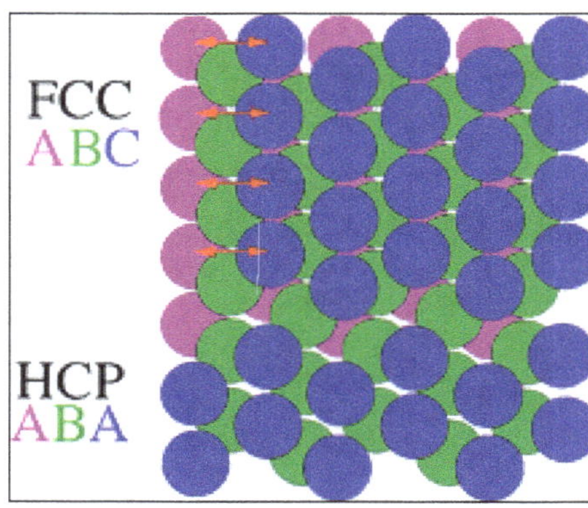

Comparison of fcc and hcp lattices, explaining the formation of stacking faults in close-packed crystals.

In crystallography, a stacking fault is a type of defect which characterizes the disordering of crystallographic planes. It is thus considered a planar defect.

The most common example of stacking faults is found in close-packed crystal structures. Face-centered cubic (fcc) structures differ from hexagonal close packed (hcp) structures only in stacking order: both structures have close-packed atomic planes with sixfold symmetry — the atoms form equilateral triangles. When stacking one of these layers on top of another, the atoms are not directly on top of one another. The first two layers are identical for hcp and fcc, and labelled AB. If the third layer is placed so that its atoms are directly above those of the first layer, the stacking will be ABA — this is the hcp structure, and it continues ABABABAB. However, there is another possible location for the third layer, such that its atoms are not above the first layer. Instead, it is the atoms in the fourth layer that are directly above the first layer. This produces the stacking ABCABCABC, which is actually along the [111] direction of a cubic crystal structure. In this context, a stacking fault is a local deviation from one of the close-packed stacking sequences to the other one. Usually, only one- two- or three-layer interruptions in the stacking sequence are referred to as stacking faults. An example for the fcc structure is the sequence ABCABABCAB.

Stacking faults carry a given formation enthalpy per unit area; this is called stacking-fault energy.

Stacking Faults in Semiconductors

Many compound semiconductors, e.g. those combining elements from groups III and V or from groups II and VI of the periodic table, crystallize in the fcc zincblende or hcp wurtzite crystal structures. In a semiconductor crystal, the fcc and hcp phases of a given material will usually have different band gap energies. As a consequence, when the crystal phase of a stacking fault has a lower band gap than the surrounding phase, it forms a quantum well, which in photoluminescence experiments leads to light emission at lower energies (longer wavelengths) than for the bulk crystal. In the opposite case (higher band gap in the stacking fault), it constitutes an energy barrier in the band structure of the crystal that can affect the current transport in semiconductor devices.

Stacking-fault Energy

The stacking-fault energy (SFE) is a materials property on a very small scale. It is noted as γ_{SFE} in units of energy per area.

The width of stacking fault is a consequence of the balance between the repulsive force between two partial dislocations on one hand and the attractive force due to the surface tension of the stacking fault on the other hand. The equilibrium width is thus partially determined by the stacking-fault energy. When the SFE is high the dissociation of a full dislocation into two partials is energetically unfavorable, and the material can deform either by dislocation glide or cross-slip. Lower SFE materials display wider stacking faults and have more difficulties for cross-slip. The SFE modifies the ability of a dislocation in a crystal to glide onto an intersecting slip plane. When the SFE is low, the mobility of dislocations in a material decreases.

Material	Ag (Silver)	Si (Silicon)	Ni (Nickel)	Cu (Copper)	Mg (Magnesium)	Al (Aluminum)
SFE (mJ m^{-2})	20-30	>42	90	70 -78	125	160-200

Stacking Faults and Stacking Fault Energy

A stacking fault is an irregularity in the planar stacking sequence of atoms in a crystal – in FCC metals the normal stacking sequence is ABCABC etc., but if a stacking fault is introduced it may introduce an irregularity such as ABCBCABC into the normal stacking sequence. These irregularities carry a certain energy which is called stacking-fault energy.

Influences on Stacking Fault Energy

Stacking fault energy is heavily influenced by a few major factors, specifically base metal, alloying metals, percent of alloy metals, and valence-electron to atom ratio.

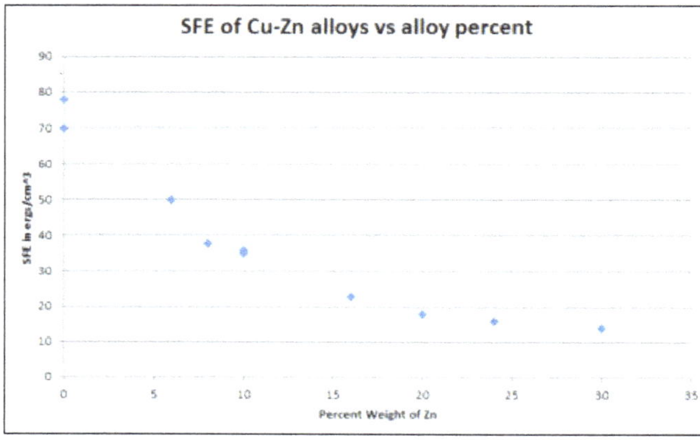

A graph of how SFE quickly diminishes with zinc alloy content.

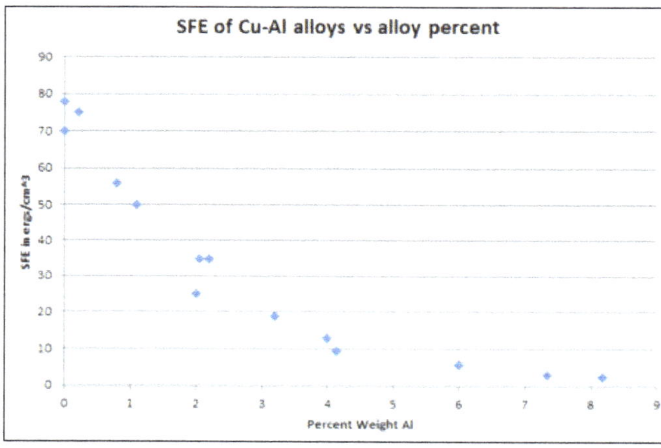

A graph of how SFE quickly diminishes with aluminum alloy content.

Alloying Elements Effects on SFE

It has long been established that the addition of alloying elements significantly lowers the SFE of most metals. Which element and how much is added dramatically affects the SFE of a material. The figures on the right show how the SFE of copper lowers with the addition of two different alloying elements; zinc and aluminum. In both cases, the SFE of the brass decreases with increasing alloy content. However, the SFE of the Cu-Al alloy decreases faster and reaches a lower minimum.

e/a Ratio

Another factor that has a significant effect on the SFE of a material and is very interrelated with alloy content is the e/a ratio, or the ratio of valence electrons to atoms. Thornton showed this in 1962 by plotting the e/a ratio vs SFE for a few Cu based alloys. He found that the valence-electron to atom ratio is a good predictor of stacking fault energy, even when the alloying element is changed. This directly supports the graphs on the right. Zinc is a heavier element and only has two valence electrons, whereas aluminum is lighter and has three valence electrons. Thus each weight percent of aluminum has a much greater impact on the SFE of the Cu-based alloy than does zinc.

Effects of Stacking Fault Energy on Deformation and Texture

The two primary methods of deformation in metals are slip and twinning. Slip occurs by dislocation glide of either screw or edge dislocations within a slip plane. Slip is by far the most common mechanism. Twinning is less common but readily occurs under some circumstances.

Twinning occurs when there are not enough slip systems to accommodate deformation and/or when the material has a very low SFE. Twins are abundant in many low SFE metals like copper alloys, but are rarely seen in high SFE metals like aluminum. In order to accommodate large strains without fracturing, there must be at least five independent and active slip systems. When cross-slip frequently occurs and certain other criteria are met, sometimes only three independent slip systems are needed for accommodating large deformations.

Because of the different deformation mechanisms in high and low SFE materials, they develop different textures.

High SFE Materials

High SFE materials deform by glide of full dislocations. Because there are no stacking faults, the screw dislocations may cross-slip. Smallman found that cross-slip happens under low stress for high SFE materials like aluminum. This gives a metal extra ductility because with cross-slip it needs only three other active slip systems to undergo large strains. This is true even when the crystal is not ideally oriented.

High SFE materials therefore do not need to change orientation in order to accommodate large deformations because of cross-slip. Some reorientation and texture development will occur as the grains move during deformation. Extensive cross-slip due to large deformation also causes some grain rotation. However, this re-orientation of grains in high SFE materials is much less prevalent than in low SFE materials.

Low SFE Materials

Low SFE materials twin and create partial dislocations. Partials form instead of screw dislocations. Screws which do exist cannot cross-slip across stacking faults, even under high stresses. Five or more slip systems must be active for large deformations to occur because of the absence of cross-slip. For both the <111> and <100> directions there are six and eight different slip systems, respectively. If loading is not applied near one of those directions, five slip systems might be active. In this case, other mechanisms must also be in place to accommodate large strains.

Low SFE materials also twin when strained. If deformation twinning is combined with regular shear deformation, the grains eventually align towards a more preferred orientation. When many different grains align a highly anisotropic texture is created.

Planar Defects

Surface defects are defects of two dimensional nature in the mathematical sense (that is, they have two dimensions). These are areas of distortions that lie about a surface, having a thickness of a few atomic diameters (that is, the thickness is almost negligible compared to other two dimensions of the surface). From the thermodynamics point of view, surface defects are not stable-and are present as metastable defects, and under suitable conditions, these defects can be removed.

Free Surfaces

Free surfaces (external surfaces) of materials are surface defects in the sense that growth of crystal has stopped abruptly at the surface. The atoms in the free surface of a solid (or, even liquid) have no neighbours and have no cohesive-bonds, on one side.

That is why, these atoms have higher free energy as compared to atoms inside. Free surfaces have surface energies that are roughly calculated from the number of 'broken bonds' at the surface. For example if the free surface is a close-packed plane of a close-packed crystal structure such as {111} type.

An atom in free surface has six neighbouring bonds in the surface plane, and three bonds with atoms below this surface but the three bonds above the surface are broken (as coordination number is 12). Thus, three out of total twelve bonds of an atom, that is, one quarter of bonds are missing compared to an atom inside.

The surface energy shall be one-quarter of the cohesive energy per atom, that is, around 1 eV for a metal such as FCC-iron. There are on an average around 10^{15} atoms cm^{-2} in a crystal surface. So, the surface energy becomes 10^{15} eV cm^{-2} i.e., 1600 ergs cm^{-2}. Table 4.8 shows experimentally measured surface free energy of some metals.

Table: Surface free energy of some materials.

Metal	Ag	Au	Cu	Fe	Ni	Pt	Sn	Si	NaCl	LiF
γ.J/m^2	1.13	1.35	1.73	1.40	1.73	2.10	0.69	1.24	0.30	0.34

Free surfaces can be classified as (a) Singular surface, (b) Vicinal surface.

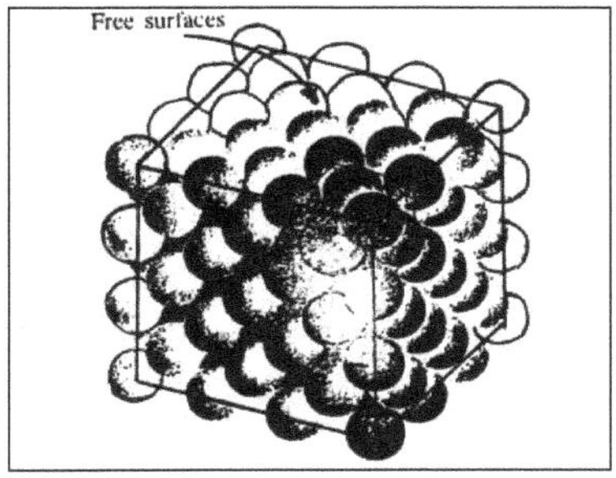

Schematic surface of a block of crystal.

Singular surface has low surface energy and usually have low index planes such as (100), (110), or (111). Vicinal surfaces have orientation near those of singular surfaces and are formed of low index planes separated by mono-atomic or mono-molecular steps or ledges, kinks and double kinks, etc.

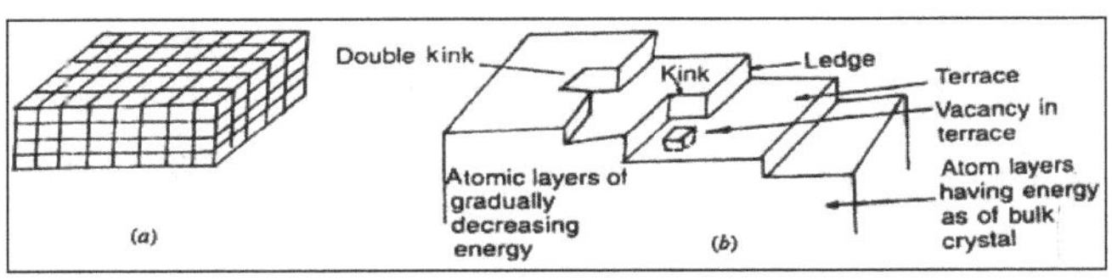

Type of free surfaces: (a) Singular surface, (b) Vicinal surface.

In metals, the free (valence) electron bond is unselective and that is why, any two metals with clean surfaces, when brought in close contact adhere well. And, it follows that a liquid metal of low cohesive energy generally spreads freely over a clean surface of a metal of high cohesive energy, because by doing so, a surface of high energy (solid metal) is replaced by a low-energy surface and a low-energy metal-metal interface. The wetting action of solders and brazing alloys is based on this principle.

Grain Boundaries

Classification of Boundaries

Dislocation theory offers good means for broad understanding of the nature and properties of grain boundaries. In a single phase polycrystalline material, a grain boundary is a two dimensional surface, which separates two regions differing in orientation of their crystal axes.

It is a surface defect where orientation of unit cells of one grain changes to other, i.e. at this surface; there is a mismatch of unit cells with unlike orientations. There are many types and gradations of grain boundaries in metals and alloys.

They are classified as:

Based on degree of difference in orientation as:

- Sub-boundaries: These are boundaries supposed to be existing within a grain, when the two regions on both sides of them have as small a difference in orientation as 1°, or less.

- Low-angle Boundaries: Low-Angle Boundaries are boundaries between two adjacent grains differing in orientation by less than about 10°.
- High-angle Boundaries: High angle boundaries are common boundaries in materials having large difference in orientation between grains.

Based on type of difference in orientation as:

- Tilt Boundaries: Tilt boundaries are boundaries between adjacent grains having misorientation in such a way that lattices appear to be tilted with respect to one another.
- Twist Boundaries: Twist boundaries are boundaries between adjacent grains having misorientation in such a way that the lattices appear to be rotated (twisted) with respect to one another.
- Coincidence Boundaries: Coincidence boundaries are boundaries in which atomic positions common to both the adjacent grains form some fraction of the total boundary area.

Based on Continuity of Atoms on both Sides of the Boundary:

- Coherent Boundary: Coherent boundary is a boundary where there is complete continuity of atoms across it, i.e. there is one to one correspondence of atoms at the boundary/interface.
- Semi-Coherent Boundary: Semi-coherent boundary is a boundary where there is partial continuity of atoms across it, i.e., some planes of atoms are continuous, but some may end at it.
- Incoherent Boundary: Incoherent boundary is a boundary having no coherency of atoms across it at all.

Sub-boundaries

When a single crystal is bent by cold working, and if one slip system operates (more dislocations are generated on one slip system), then the bent crystal looks as in figure (a), because excess positive edge dislocations remain in the crystal to accommodate the curved shape of the crystal.

When such a bent crystal is annealed (before the recrystallisation takes place), the curved crystal forms a number of strain-free sub-grains, each preserving the local orientation of the original bent crystal separated by plane sub-boundaries. This process of forming sub-grains and sub-boundaries is called polygonisation, because a smooth curved crystal changes into polygons.

The dislocations have arranged themselves in sub-boundaries (by the processes of slip and climb during annealing), one above the other because in this position, dislocations relieve each other's elastic strain fields, figure (d). The dislocation arrangement of figure (a) is one of high strain energy, which changes to as in figure (b) of lower strain energy.

The sub-boundaries between the sub-grains are considered to exist within a grain and thus, the difference of orientations between the two adjacent sub-grains is very small, around 1° or less. Sub-boundaries are small-angle or symmetric tilt-boundaries within the grain having very small difference of orientation between the sub-grains, figure (e). Sub-boundaries are revealed by etching as dense row of pits under microscope.

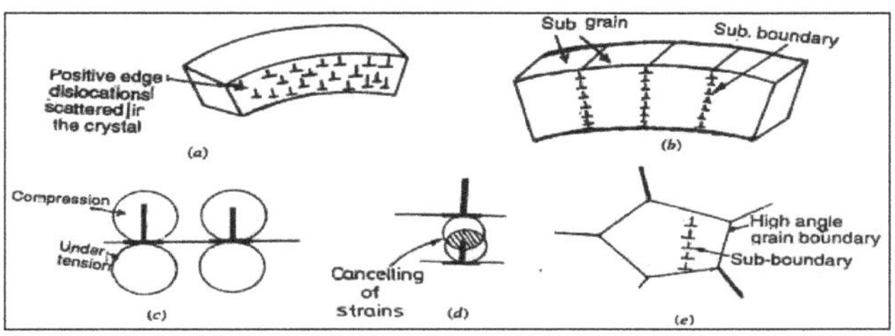

Formation of sub-boundaries, (a) A bent cold worked single crystal with excess dislocations of one sign (here positive edge dislocations), (b) After polygonisation, sub-boundaries are formed within the crystal itself, (c) Dislocation on same plane, (d) Tensile strain of upper dislocation is partly cancelled by compressive strains of lower dislocation. (e) Schematic representation of array of parallel positive edge dislocations within a grain at a sub-boundary.

Low-angle Boundaries or Tilt-boundaries

In 1940, both Bragg and Burger, and later Read-Shockley proposed the dislocation model of structure of low-angle boundaries with good success. Low angle boundaries are boundaries between adjacent crystals of same structure but with small difference in crystal orientations, and might be considered to have an array of parallel dislocations (along the plane of the boundary).

Two grains in schematic figure (a) are misoriented or pictured as tilted symmetrically about an axis (horizontal inside the paper), parallel to the direction of edge dislocations. The two lattices can be maintained continuous by periodic addition of an edge dislocation.

The resulting vertical array of parallel dislocations of same sign (here, positial edge dislocations) is just like a sub-boundary but with greater difference of orientation, still less than about 10°, and is in between two adjacent grains. Low-angle boundaries are also observed under microscope as uniformly spaced line of etch pits, figure (c).

The presence of each dislocation at the surface of a specimen as etch pit is revealed because of the faster rate of chemical attack by the etchant in the elastically strained area of the crystal near the core of the dislocation.

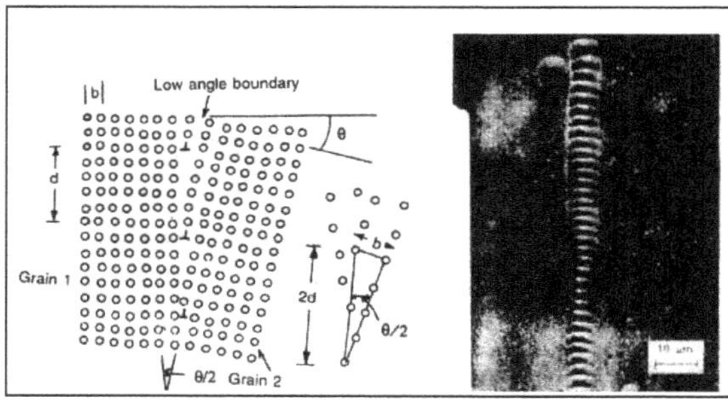

Nature of low-angle boundary (tilt boundary) in metals: (a) Dislocation model of low-angle gain boundary, (b) Relationship between θ, the angle of tilt with d, the spacing between two dislocations, (c) Etch pits seen under microscope revealing the low angle boundary having regularly spaced dislocations.

The angle theta is the angle of misorientation. The greater is the angle of misorientation, the greater is the inclination of planes that terminate as dislocations at the boundary, and closer is the spacing of the dislocations in the vertical boundary, i.e., the spacing between the dislocations in the boundary determines the theta, the angular inclination between the lattices. In figure (b), we find that:

$$\sin\theta/2 = b/2d$$

where, b is the burgers vector of dislocation in the boundary, and d is the spacing between the two adjacent dislocations. If the angle of misorientation is assumed to be small, then $\sin\theta/2 \approx \theta/2$, and,

$$\theta = \frac{b}{d}$$

and the results of this relationship have been found to be in good agreement by x-ray diffraction determinations, and confirms the validity of the dislocation model for low-angle boundaries.

This relationship could be used to drive the energy of the boundary as dependent on angle of misorientation. With the increase in misorientation angle, θ, of a symmetric tilt boundary, the value of d decreases, i.e., the number of dislocations increases.

We know by now that total energy of each edge dislocation per unit length is given by:

$$E_b = E_C + \frac{Gb^2}{4\pi(1-v)} \text{In} \frac{d}{2r_o}$$

where, Ec is the energy per unit length of dislocation core. Energy of the dislocation decreases as d becomes small, or θ increases ($\because d = b/\theta$). If we assume $r_o = 2b$, the total energy of the dislocation per unit length:

$$E_b = E_C + \frac{Gb^2}{4\pi(1-v)} \text{In} \frac{1}{4\theta}$$

If 'd' is distance between two dislocations, then energy of unit area of the boundary is approximated as:

$$E_b = \frac{1}{d}\left[E_C + \frac{Gb^2}{4\pi(1-v)} \text{In}\frac{1}{4\theta}\right]$$

$$= \frac{\theta}{b}\left[E_C + \frac{Gb^2}{4\pi(1-v)} \text{In}\frac{1}{4\theta}\right] = \theta\left[\frac{E_C}{b} + \frac{Gb}{4\pi(1-v)} \text{In}\frac{1}{4\theta}\right]$$

$$= \theta\left[\frac{E_C}{b} - \frac{Gb}{4\pi(1-v)} \text{In}\,4 - \frac{Gb}{4\pi(1-v)} \text{In}\,\theta\right]$$

$$= \theta[A - B\,\text{In}\,\theta]$$

where,

$$A = \frac{E_C}{b} - B\,\text{In}\,4, \text{and}$$

$$B = \frac{G_b}{4\pi(1-v)}$$

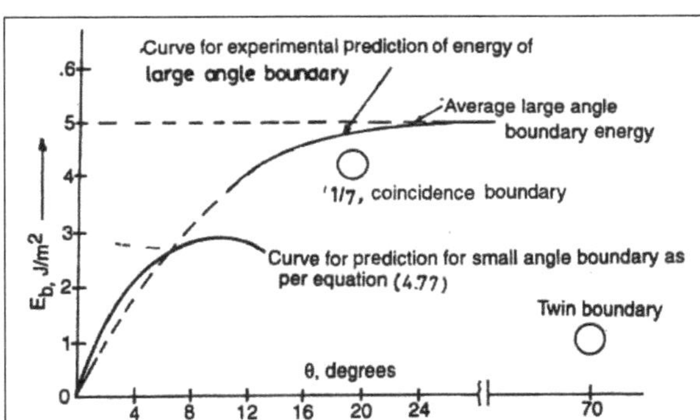

Comparison of energy for experimental data and theoretical data for copper.

If a graph is plotted between E_b and θ as in figure and experimental data is superimposed for copper, the curve matches well up to below about 8°. There is no maximum obtained experimentally. The equation above predicts well that as b and E_C approach zero, the energy vanishes.

At higher angles than 8°, the dislocations are quite closely spaced, and are difficult to be distinguished due to overlap of cores of the dislocations. The structure of the grain boundaries beyond 8° of misorientation appears to acquire more complicated configurations. This model of dislocations appears to be successful for low angle boundaries up to about 8°.

Problem: Copper shows array of edge dislocation an (110) plane of atoms. Calculate the spacing of dislocations if the angle of tilt is 1°. (Lattice constant of copper 3.62 A).

Solution: Copper is FCC, and its (110) plane is shown on which dislocations have Burgers vector:

$$\therefore \text{ Burger vector } = \frac{3.62}{\sqrt{2}}$$

$$= 2.56 \text{ A}$$

As $\theta = 1° = \frac{1 \times \pi}{180}$ radians.

$\therefore$ d = distance between two neighbouring dislocations

$$= 2.56 \times \frac{180}{\pi}$$

$$= 146.70 \text{ A°}$$

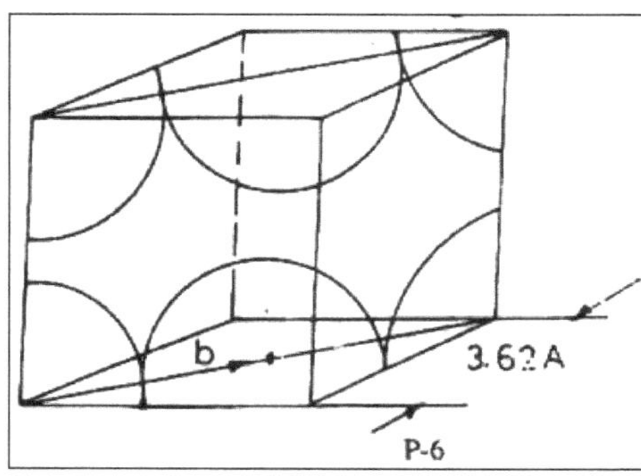

Degrees of Freedom of Grain Boundary

All single phase poly-crystalline materials contain a complex network of grain boundaries between adjacent grains that are crystallographically misoriented with respect to one another. The grain boundary where the two neighbouring grains appear to be inclined or tilted with each other (by a small angle) about a horizontal axis normal to the paper is a very special boundary, as it has only one degree of freedom, or variable. In fact, the grain boundaries have five degrees of freedom, three to define the rotation, and two to define the boundary plane.

We specify the difference in orientation by the rotations which we would have to give to one crystal to bring it into same orientation as the other crystal. For example, if we rotate the grain 1 in figure (a) by an angle theta clockwise about a horizontal axis which runs inside normal to the paper, the two grains become in the same orientation.

There are thus, three degrees of freedom available for such rotations, i.e. there are three ways of doing it corresponding to the three perpendicular axes about which rotations can be made. There are two more degrees of freedom for choosing the orientation of the grain boundary itself, that is, the boundary may not be in symmetrical position between the two crystals, but can be rotated about either of the two axes at 90° to each other.

We can tilt or twist one crystal relative to the other in three ways, and have two ways to align the boundary between the crystals. Figure illustrates schematically these five degrees of freedom of a grain boundary. In figure (a), the two neighbouring crystals are tilted with respect to each other about a horizontal axis, normal to the plane of paper with grain boundary placed symmetrically.

This boundary has one degree of freedom and is a symmetrical tilt boundary. Figure (b) illustrates also a symmetric tilt boundary but with a vertical tilt axis. In both these cases, the rotating tilt axis lies in the boundary. The boundary has edge dislocations in it. In figure (c), one crystal has been rotated relative to the other about a horizontal axis perpendicular to the boundary.

This is called a symmetric twist boundary, and normally contains at least, two different arrays of screw dislocations. Figure (d) and (e) illustrate that the grain boundaries do not have to be in symmetric positions between the two crystals, but may align asymmetrically in two different ways between the crystals—one by rotating the boundary along a horizontal axis, figure (d) and other along a vertical axis.

Majority of the grain boundaries in single phase polycrystalline materials have a mixed character of these five variables of varying degree, and thus the general or large angle boundaries have a complex structure.

Five degrees of freedom of a grain boundary: (a) Symmetric tilt boundary with horizontal axis of tilt (b) Symmetric tilt boundary with vertical tilt axis, (c) Symmetric twist boundary, (d) Asymmetric tilt boundary, (e) Asymmetric tilt boundary with vertical axis of boundary rotation.

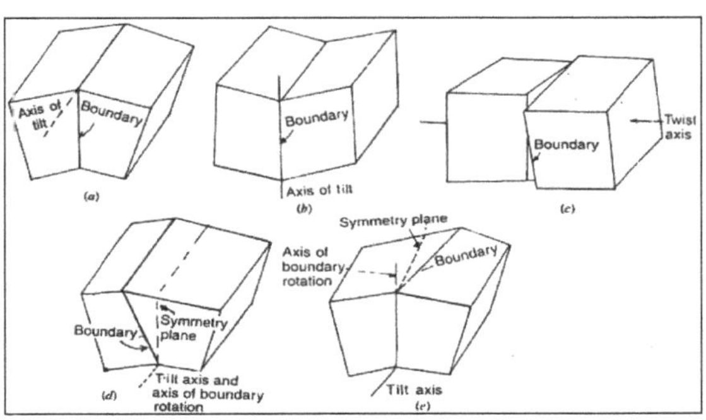

A symmetric low angle boundary, or symmetric tilt boundary, figure has angle of misorientation equally divided between the two neighbouring grains, i.e., each lattice is displaced by $\theta/2$ from the vertical axis, i.e., has one degree of freedom. But a typical tilt boundary may have two degrees of freedom (or even more), such as asymmetric tilt boundary, figure (a), where the angle of misorientation is not divided equally between the grain 1 and grain 2.

Let us take a special case, figure (a), where the angle has been so chosen that grain 1 contributes twice as many edge dislocations of Burgers vector $\vec{b_1}$, as by grain 2 of

Burgers vector, b2→. Let us assume that 2 dislocations of grain 1 combine with one dislocation of grain 2, by reaction,

$$2\,b1^{\rightarrow} + b2^{\rightarrow} = b3^{\rightarrow},$$

to give a resultant average dislocation of Burgers vector, Although this reaction does not occur, but it can be assumed that these dislocations of Burgers vector, lie along the asymmetric tilt boundary, and is the Burgers vector perpendicular to such an asymmetric boundary.

Depending on the position of the asymmetric tilt boundaries, the angle θ of misorientation may be shared in innumerable ways between the grains to give an average resultant dislocation of different Burgers vectors. Thus, a large number of such asymmetric boundaries can be assumed to exist.

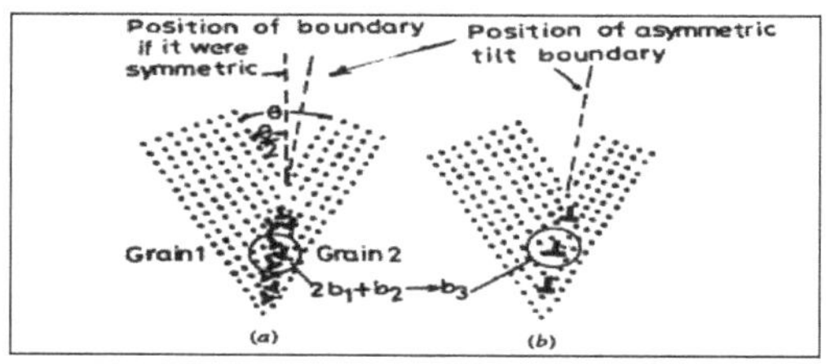

Asymmetric tilt boundary with two degrees of freedom, (a) Asymmetric tilt boundary. As the boundary has been rotated along horizontal axis, grain 1 adds twice as many dislocations as grain 2. (b) same as (a), but 2 dislocations of grain 1 and 1 of grain 2 result to give an equivalent average dislocation $\left(2b1^{\rightarrow}+b2^{\rightarrow}+b3^{\rightarrow}\right), b_3$, which are assumed to be lying along the boundary.

Twist Boundary

A small-angle pure twist boundary is obtained, if one of the crystals appears to be rotated or twisted relative to its neighbour about an axis perpendicular to the plane of the boundary, figure.

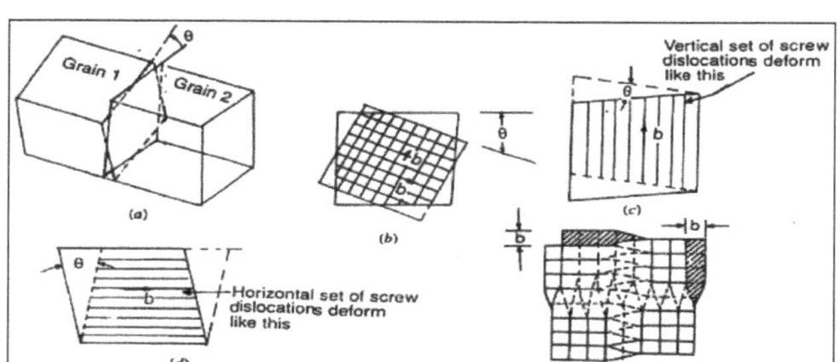

Twist boundary: (a) Grain 2 has been twisted with respect to grain 1 along an axis perpendicular to boundary, (b) Both type of screw dislocations, that is, crossed grid of screw dislocations causes the twist to permit a transition between grain 1 and 2; (c)

Shape of cross-section at slip plane of a square crystal deformed by a set of vertical screw dislocations; (d) by a set of horizontal screw dislocations; (e) Two intersecting screw dislocations of (d) and (c).

We see shape of cross section at the slip plane of a square crystal lattice deformed by a set of vertical uniformly spaced screw dislocations. The two surfaces on either side of the slip plane have been displaced by an angle, θ, but such a set of vertical dislocations have not accomplished the desired rotation needed to get a twist boundary of figure (a).

If only a horizontal set of screw dislocations are introduced, then the result is illustrated in figure (d), which also does not match with figure (a). If simultaneous introduction of two sets of screw dislocations is done, we get rotation, or twist of two crystals relative to each other to get a picture.

A twist boundary, lying along a crystallographic plane has rotational displacements of atoms symmetric with respect to the boundary. This type of boundary requires crossed grid of screw dislocations to permit continuity from grain 1 to grain 2. If in a unit area of boundary, each set of screw dislocations contain n dislocations, then the twist angle, θ, is:

$$\theta = n\vec{b}$$

If the twist boundary does not lie along a crystallographic plane, then, it is asymmetric with respect to rows and columns of atoms on either side of it. At least, then, three sets of screw dislocations with different Burgers vector are needed to produce an asymmetric twist boundary.

An asymmetric twist boundary is analogous to asymmetric tilt boundary. The actual small angle boundary of arbitrary orientation may have varying five degrees of freedom that is, varying degree of tilt and twist of crystallographic axis. If the boundary is asymmetric, then, it may, contain edge, screw and mixed dislocations.

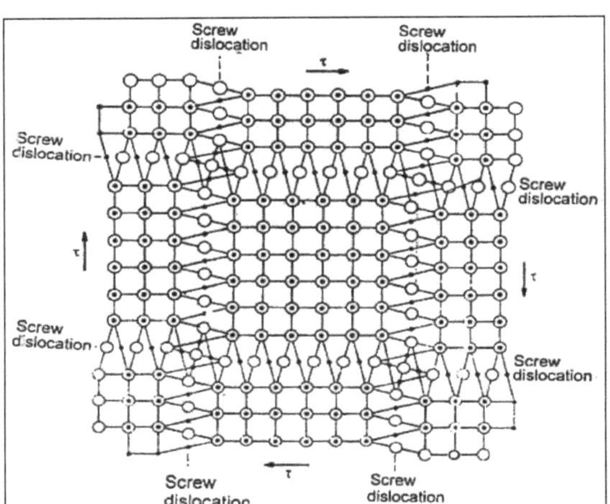

Symmetric twist boundary between between two crystals composed of two sets of intersecting screw dislocations in a simple cubic lattice.

Coincidence Boundary

If we make duplicate square lattice of holes in a black paper, and then pin the two sheets together, say at X in figure (a) and hold them against light and rotate into positions in which various other points come into coincidence. One such seen is as in figure (a). figure (b) illustrates a square lattice.

A rotation of 37° about a lattice point X, makes atom Y to coincide with Y', and Z to Z' in the un-rotated lattice. The figure (c) illustrates two neighbouring grains in which grain 2 has been shown after rotation of 37° relative to grain 1. Dark dots are coincidence points.

It is clear that coincidence sites are continuous in both the grains in spite of grain boundary between them. These coincidence sites appear to form a lattice of themselves (joined by fine lines), but of larger cell dimensions. We see that one atom out of five atoms is a coincidence site, so that the term 1/5 coincidence boundary is applied to this arrangement of points.

Table illustrates some coincidence lattices in BCC and FCC metals. Coincidence lattices occur in all of the common crystal structures, and have site densities varying from 1/3 to 1/19 and even lower. Coincidence boundary normally is incoherent boundary like an ordinary grain boundary; that is, most of the atoms at the boundary of one crystal do not match the lattice positions in the other crystal, but mismatch is less, and so have less energy than ordinary boundaries.

Table: Some Coincidence lattices in FCC and BCC structures.

Structure	Axis of rotation	Angle of rotation	Density of coincidence sites
FCC		22.0	1/7
		38.2	1/7
		38.9	1/9
		36.9	1/5
BCC		60.0	1/3
		38.9	1/9
		70.5	1/3
		36.9	1/5

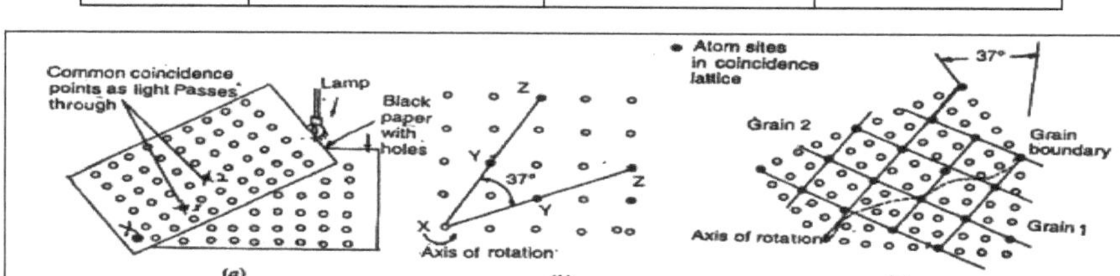

Coincidence lattice in square lattice (two dimensional). (a) Two black card-board with holes of square lattice and both pinned at X and one is rotated keeping the other fixed and seen against a lamp. Two coincidence points, 1 and 2 are seen, (b) A square lattice when rotated about its origin X by 37° make atom Y to coincide with Y' and Z to Z', (c) When two adjacent square lattices are misoriented by 37°. One atom out of five is a co-incidence site, i.e., $1/5$ coincidence lattice exists.

As coincidence sites are low energy sites, the grain boundary between the grains may not follow a relatively straight path as in figure (c) but a zig-zag path of figure (a), so that the grain boundary has high density of coincidence sites. Most coincidence lattices do not have high densities of coincidence sites, but in certain cases, the coincidence sites lie in moderately close-packed planes as in figure (b), leading to the formation of coherent boundary.

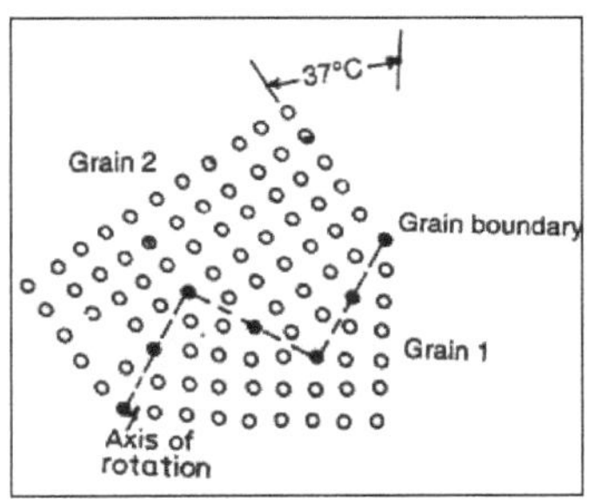

Grain boundary takes a zig-zag path to pass as many coincidence sites as possible.

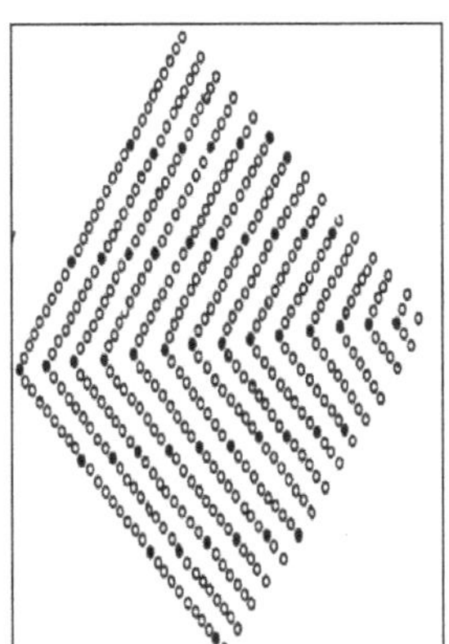

In BCC lattice, the neighbouring (210) planes obtained after a rotation of $94^{\circ}-4'$. A high density coincidence boundary which is also a coherent boundary.

Coherent, Semi-coherent and Incoherent Boundaries

A coherent boundary has all the atom sites in it common to both crystals (regions) joined by it. There is, thus, continuity of crystal structure from one grain to the other, or as we say, there is one to one correspondence of atoms at the boundary. As the atoms in the boundaries fit together well, coherent boundaries have low energies, and that is why all boundaries tend to become as coherent as possible.

Twin boundary is a coherent boundary of figure (a). Other examples of such boundaries are in widmanstatten structures, martensite, and in precipitates from supersaturated solid solutions. The important requirement for the formation of coherent boundary is that the atoms in the adjoining crystals must, at the common boundary (coherent), have closely similar patterns and spacings.

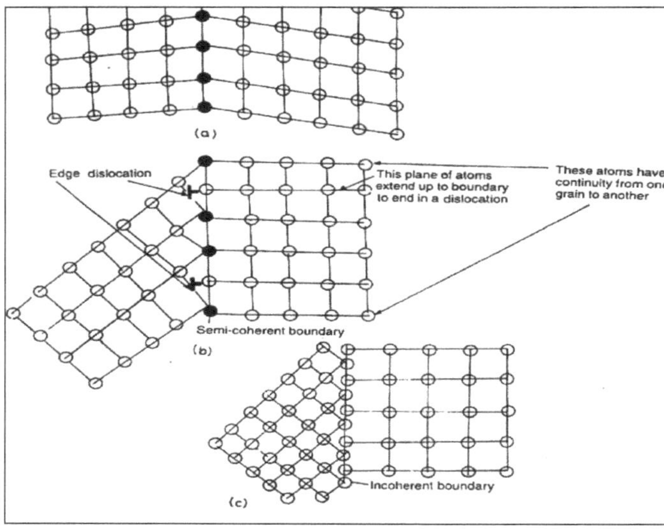

Schematic illustrations of coherent, semi-coherent and incoherent boundaries in a single phase poly-crystalline material. (a) Coherent boundary in a single phase poly-crystalline materials, (b) Semi-coherent boundary in a single phase poly-crystalline material, (c) Incoherent boundary.

Small-angle boundary is a semi-coherent boundary. If the dislocations get introduced in a coherent boundary, it results in a semi-coherent boundary. There is continuity of atoms in some layers in both the adjacent grains, but ends up in a dislocation in other layers at the boundary.

By suitably and steadily increasing the density of dislocations in a boundary, (to represent increasing rotations of the crystals, or the boundary, or both) from orientation that produces coherent boundaries, a gradual change from coherent to semi-coherent to in-coherent boundaries can be obtained, figure illustrate this change. In the last extreme limit, the dislocations are too densely packed to overlap cores of neighbouring dislocations, to result in incoherent, or so called normal grain boundaries.

In multi-phase materials, when second phase forms in alloy crystals, then to get a coherent interface between the phases, the crystal structures of both phases have to be matched. If the size of precipitate is very small, it may adopt a crystal structure that allows it to be coherent with the matrix crystal, even if for matching the atomic bonds of two crystal structures at the boundary produce considerable elastic distortions in it, and its immediate surroundings.

The lattice constant of one phase increases and the other decreases. The thin disc shaped particles have least strain energy, and still be coherent. These thin tiny particles, which are coherent, can be thought of as solute-rich clusters in the parent matrix. In quenched and aged aluminium with 4 percent copper alloy, the tiny particles are copper rich discs about 100 A in diameter and 4 A thick on {100} planes of aluminium matrix with coherent interfaces, figure (a).

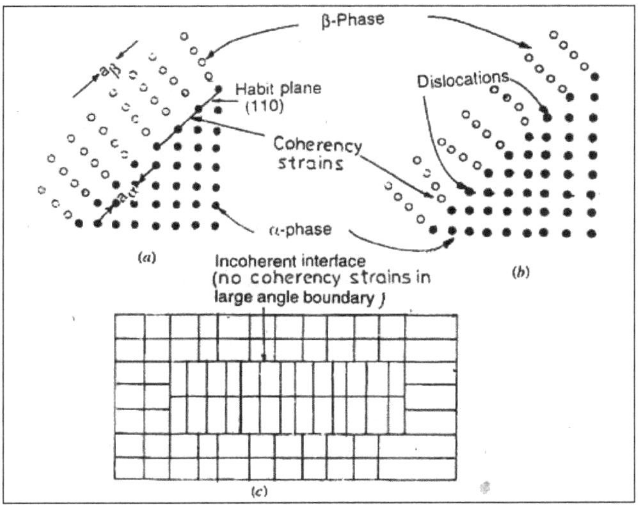

Coherent, semi-coherent and incoherent interfaces (boundaries) in poly-phase materials. (a) A coherent interface with lattice strain due to the disregistery, $\delta = \frac{a\beta - a\alpha}{a\beta} < 0.1$, (b) A semi-coherent interface with dislocations, which partially relieve the lattice strains due to registry, (c) Incoherent interface (no lattice strains in a large angle boundary).

When such a coherent particle grows larger by obtaining atoms from matrix and crystal structure changes, it may break away (may be partly in the beginning) from its coherent attachments to the matrix in order to reduce the increasing elastic coherency strains, and adopts a more stable structure by introducing dislocations in its interface to make the interface semi-coherent, which on further change of the precipitate forms an incoherent interface.

Common Grain Boundaries

These boundaries are formed when two neighbouring grains, solidifying from liquid metal, or recrystallising, impinge on each other. The resulting boundary can be quite arbitrary because there are five degrees of freedom of such a boundary. The dislocation model of low-angle grain boundary illustrates that with the increase in angle of misorientation between the grains, the dislocations lie closer together.

At sufficiently high angles of misorientation, the individual dislocations lose their identity, because the cores of the dislocations overlap. The structure and the energy of such boundaries, called the large angle boundaries or common boundaries are difficult to be predicted. However, it is clear that grains on both sides of them are truly crystalline up to the boundary, which is a narrow transition region of imperfect order.

The grain boundaries are not necessarily smooth curvilinear surfaces, but with ledges and facets of macro as well as microscopic dimensions, particularly if the grains are held at-sufficiently high temperature to minimise the local elastic strains, figure.

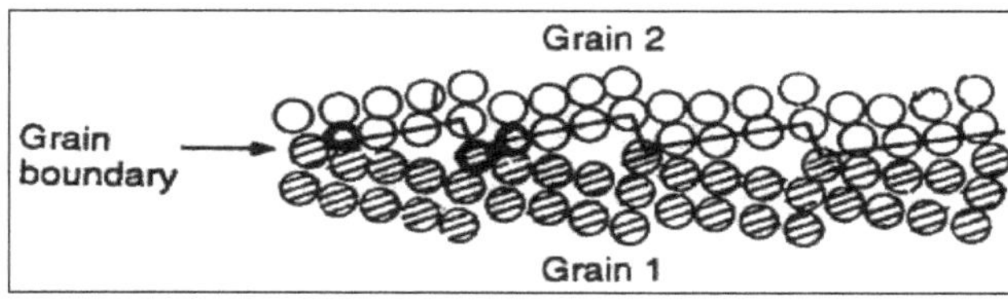

Ledges and facets in a grain boundary.

The energies of large-angle boundaries are usually estimated by thermal-etching. When a polycrystalline material is heated at high temperatures, thermal grooves form on the free surface, wherever a grain boundary intersects the surface perpendicularly.

This is an example of general effect of balancing the forces of surface tension at an interfacial triple junction, figure (a), consisting of one grain boundary and free surfaces to the right and to the left of the point X.

If Υ_s is the free surface tension and Υ_b is surface energy of the boundary, then the two surface tensions of the free surface balance each other, but the vertically down component due to grain boundary remains unbalanced.

In order to balance this grain boundary energy, a groove must form as illustrated, figure (b) with a dihedral angle, θ between the two surface energies of free surfaces to satisfy the equation, figure (c):

$$\Upsilon_b = 2\Upsilon_s \cos\theta/2$$

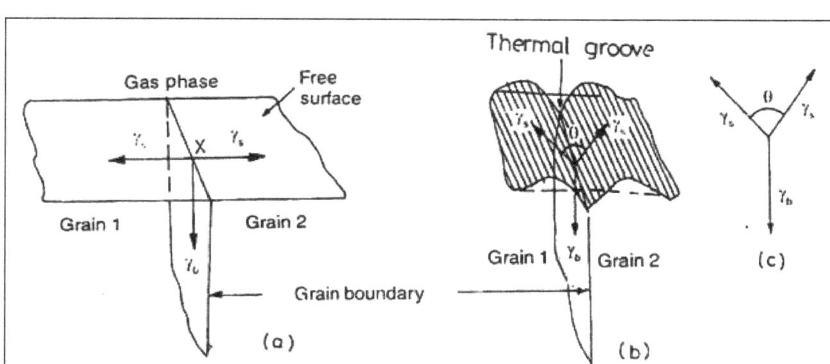

Schematic illustration of formation of thermal groove. (a) A grain boundary meets a free surface perpendicularly. (b) Thermal groove is a stable configuration at the junction of grain boundary with the surface, (c) Interfacial triple junction at groove base.

Diffusion and evaporation processes at high temperatures enable the necessary boundary movements to take place. It becomes possible to estimate Υ_b if θ and Υ_s are known. Actually, this equation gives free energy of a boundary, because the total energy is larger than this, due to extra contribution associated mainly with the vibrational entropy of atoms in the boundary. Table provides free energies of some large-angle boundaries. We find, $\Upsilon_b \approx 1/3\ \Upsilon_s$.

Table: free energies of large-angle boundaries of some metals.

$\Upsilon_{b,}$ Jm^{-2}	Au	Cu	: Fe	Ni	Sn
	0.36	0.60	0.78	0.69	0.16

Grain boundaries have dislocations and vacancies in them. Infact, grain boundaries are source and sink of them. Soluble impurities strongly segregate to the grain boundaries to reduce their elastic misfit energies In fact, first few atomic percent of impurity as well as alloy atoms are invariably present at the grain boundaries as these can be accommodated better in imperfectly ordered grain boundaries.

In zone-refined metals, grain boundaries migrate readily at around 10^{-5}m sec^{-1} at temperatures close to the melting point, and some movement has been also detected at 0.25 $T_{m.p.}$. But soluble impurities at the grain boundaries decrease the movement of grain boundaries. The insoluble impurities have an even stronger effect in reducing grain boundary motion (also thus grain growth).

Problem: A thermal groove was observed on the surface of iron where the grain boundary intercepted it perpendicularly making a dihedral angle of 147°. The surface energy of iron is 1.49 Jm^{-2}. What is the grain boundary energy of iron?

Solution:

$$\begin{aligned} \Upsilon_b &= 2\Upsilon_s \cos\theta/2 \\ &= 2 \times 1.49 \times \cos 73.5^\circ \\ &= 0.85\,\mathrm{Jm}^{-2} \end{aligned}$$

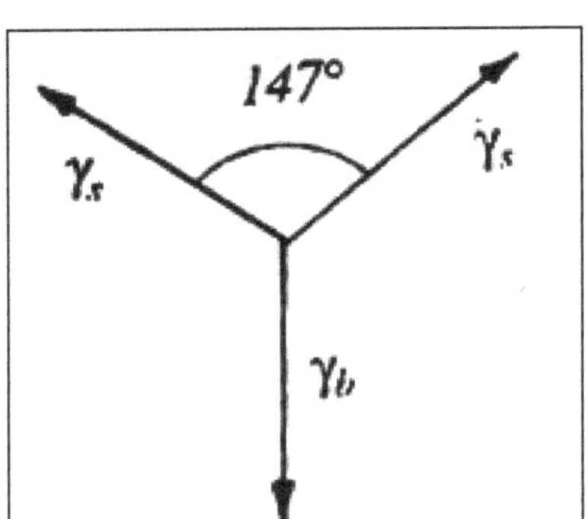

Interfacial triple junction at thermal groove.

Stacking-faults

Stacking-faults are planar surface defects produced by a fault (mistake) in the stacking sequence of atomic planes in crystals. The stacking sequence of planes of atoms on {111} planes in FCC crystal structure has been designated as ABCABC..., whereas in HCP crystal structure on (0001) plane as ABABAB....

A fault in stacking sequence of say FCC crystal structure may be produced by plastic deformation, or by collapse of a large number of vacancies on one plane, or by agglomeration of a large number of interstitialies on a plane, etc.

Suppose a large number of vacancies collapse on a (111) plane, so that one plane of atoms is missing, say plane C is missing indicated by arrow:

$$\ldots ABC\,AB\overset{\downarrow}{C}\,ABC\ldots,$$

then the stacking sequence becomes:

... ABC AB ABC ...

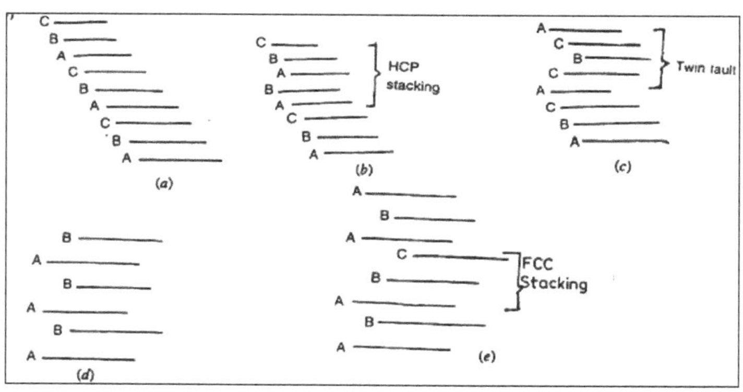

Stacking faults in FCC and HCP crystals: (a) FCC stacking sequence ABC ABCABC, (b) Deformation fault in FCC: ABCAB:ABC, (c) Twin fault in FCC: ABCACBCABC (d) HCP stacking sequence, (e) FCC fault in HCP crystal.

There is now a fault in stacking, and this type of fault is called intrinsic stacking-fault as also seen in the formation of Frank loop by the internal collapse of vacancies. If we compare this sequence (underlined stacking sequence) with an HCP stacking sequence without any fault, that is, ... AB AB AB... figure (d), we find that stacking-fault contains four layers of an HCP stacking sequence.

Thus, formation of an intrinsic stacking-fault in an FCC metal is equivalent to the formation of a thin HCP region. This thin region is a surface defect called stacking-fault. The creation of a region with HCP stacking ABAB introduces a region with higher free energy than the FCC structure.

The coordination number in the faulted region is though twelve as in the unfaulted or perfect region, but the second-nearest-neighbour bonds in the faulted region are not of correct type for FCC perfect lattice. Thus, energy is associated for the creation of the stacking-fault, and is called stacking fault energy, which is normally of the order of 0.01 to 0.05 Jm^{-2}.

Table: Stacking fault width and energies of some materials.

Metal	Stacking fault energy, ϒ J/m²	Equilibrium width of faulted region, d_g, A
Lead	0.025	-
Silver	0.025	18.7
Gold	0.050	5.7
Copper	0.080	6.8
Aluminium	0.200	1.5 (considered unextended)
Nickel	0.240	2.0
Brass	0.01	-
Stainless steel 303	0.008	-

If some interstitialies agglomerate on a plane of atoms, addition of a plane of atom (at least in some area of it) takes place, It is because of this area of extra atoms as a part of plane that the stacking sequence is faulted, and such a stacking fault is called an extrinsic stacking fault. For example, HCP crystal having a stacking sequence of:

$$ABAB\overset{\downarrow}{A}B\ ...$$

is made to have an extra plane of atoms added, say, at the place of the arrow, then the atoms of this extra plane occupy a position different than positions of A or B, but say at C. At positions either A or B, the structure shall be highly distorted as then we shall have either position B on B or A on A. Now, at C positions, the stacking sequence becomes:

$$...AB\underline{ABC}AB\ ...$$

that is, an HCP crystal has a stacking-fault which contains three layers of FCC stacking sequence, and thus it shall require extra energy. Because of the additional lattice discontinuity associated with an extrinsic fault, it has higher energy than an intrinsic fault.

The introduction of an extrinsic fault in FCC crystal structure changes the stacking sequence from:

$$ABCA\overset{\downarrow}{B}CABC...$$

to ABCACBCABC ...

By adding an extra plane of atoms at C positions in between A and B as indicated by arrow. The three layers ACB constitute a twin (of BCA). Thus, an extrinsic stacking-fault in FCC crystal is also called twin stacking fault.

This stacking fault in FCC metals can also be considered sub-microscopic twin of approximate atomic thickness. The stacking-fault energies of different FCC metals is responsible for their differential deformation behaviour. Metals in general with low-stacking-fault energies strain harden more rapidly, and twin easily on annealing.

Stacking-faults are easily detected after etching under an optical microscope, because of difference in the etching characteristics of faulted and parent structure.

Topological Defect

A topological soliton occurs when two adjoining structures or spaces are in some way "out of phase" with each other in ways that make a seamless transition between them impossible. One of the simplest and most commonplace examples of a topological soliton occurs in old-fashioned coiled telephone handset cords, which are usually coiled clockwise. Years of picking up the handset can end up coiling parts of the cord in the opposite counterclockwise direction, and when this happens there will be a distinctive larger loop that separates the two directions of coiling. This odd looking transition loop, which is neither clockwise nor counterclockwise, is an excellent example of a topological soliton. No matter how complex the context, anything that qualifies as a topological soliton must at some level exhibit this same simple issue of reconciliation seen in the twisted phone cord example.

Topological solitons arise with ease when creating the crystalline semiconductors used in modern electronics, and in that context their effects are almost always deleterious. For this reason such crystal transitions are called topological defects. However, this mostly solid-state terminology distracts from the rich and intriguing mathematical properties of such boundary regions. Thus for most non-solid-state contexts the more positive and mathematically rich phrase "topological soliton" is preferable.

In mathematics and physics, a topological soliton or a topological defect is a solution of a system of partial differential equations or of a quantum field theory homotopically distinct from the vacuum solution.

Conservation Laws for Crystal of Topological Defects

Topological defect arises when the symmetry of order in material is broken. The symmetry breaking can be induced by phase transitions or by application of confinements. The types of defects, such as point, line, or walls, depends on dimension of the order and topology of the confinement. Though topological defects represent the corruption of the previous order, they trap molecules and particles. The network between defects provide a scaffold, or the attraction between defects triggers the self-assembly and holds the inclusions together. Periodic, regular, stable topological defect network mediates the birth of a new order. The dynamics and mechanism of defect network formation explain why and how colloids biological fibers and tissues grow into a particular structure.

Two-dimensional (2D) colloidal crystal assembled by attraction between topological defects in nematic liquid crystal (NLC) was discovered in 2006. The "boojum" point defect attached on the particle generates a dipolar director field. The positive and negative poles of the dipoles attract each other, and the particles line up in a chain. The chains attract each other when the dipoles are anti-parallel, forming adipolar crystal by self-assembly. The "Saturn ring" loop defect around the waist of the particle generates a quadrupolar direct or field, resulting in the hexagonal crystals. Three-dimensional (3D) dipolar crystal was successfully assembled in 2013. Since 2015, pure topological defect arrays without inclusions have been generated by using crossed strip electrodes, patterned surface alignment micro cavities on wafer, doping of ions in NLC, standing pressure wave, and interaction between optical vortex and liquid crystal. These methods set up periodic confinements, and the symmetry breaking of NLC in periodic confinements leads to array of umbilical escapes or disclinations. The defect arrays are topologically protected and self-retained. Therefore, they are promising templates for wires, fibers, or tissues.

In analogy to mineral crystals, defectarray has periodicpatternthatexhibit long-range order. One topological defect (or a group of topological defects) can be one unit cell, and the periodic repetition gives a crystal. In mineral crystals, the electron configuration of atoms determines the rotational symmetry of the unit cell, and the periodic repetitionof unit cells renders the rotational symmetry of the crystal. In NLC, the symmetry of isotropic distribution is brokenby the rod-like shape of molecules, leading to long-range nematic order. Then the symmetry of nematic order is broken by periodic confinements again, leading to topological defect crystals having "extra" long-range order. There are two hierarchies of orders: the nematic order and the periodic arrangement of defects. The shape of the topological defect, such as radial, circular, hyperbolic hedgehogs, lines, walls, which are associated with topological charges, must match to the periodicity of the defect crystal. The director field across the boundary must be continuous, except the locations of singularities. The strain between the defect must have static balance, so the crystal of topological defects can be stable.Based on the above mentioned principles, selection rules of stable topological defect array can be developed. Above the selection rules, conservation laws of director field may be hidden in the structure of topological defect array. To observe and analyze the defect crystal systematically, topological defects were generated in pixelized NLC test cell and arranged on square or hexagonal arrays. Forthe first time,large arrays of points defect with designed shapes were created. The distinction between stable and unstable crystals were clear. Based onthe topology and symmetries of the director field, conservation laws of director flux and topological charges were discovered.

Structure of the Liquid Crystal Cell

Array of topological defects was generated in test cells with homeotropic alignment on the surface. NLC with negative dielectric anisotropy (Δε)of −3.0was sandwiched between glass substrates with Indium-Tin-Oxide (ITO) sputtered on the surface. The patterns of ITO were made of lithography and wet etching. Polyimide (PI) was coated on the glass substrate, so the NLC stood vertically on the surfaces. The bottom substrate was patterned, and the top substrates was a full ITO. NLC was injected into the test cell by capillary effects. Square wave AC voltage with frequency of 60 Hz was applied on the bottom ITO, and top ITO was connected to the ground. The pixel array was patterned on the bottom substrate. The pattern of individual pixels could bea pad, a fishbone, or a coil. Then the individual pixels were arranged in square or hexagonal lattices. The topological defects generated by the pixels were be arranged as if a 2D crystal, and each pixel was a unit cell.When the voltage is applied, the ordinary axis of negative NLC align along the direction of the electric field, which means that the vertically standing NLCwill recline and lie in plane (the x-y plane). In a pixel as shown in figure, NLC on the edge of the bottom electrode reclines and points into the center of the pixel due to strong fringe field. The vertically aligned NLC on the "celling (top)", on the "floor (bottom)",and the reclined NLC on the "walls" compose a "box" with homeotropic surfaces. The pixel becomes a closed homeotropic boundary when electric field is applied. A homeotropic box is topologically equivalent to a homeotropic sphere, inwhich the liquid crystal configuration can be anumbilical escape (ubmilic) or a point defect.

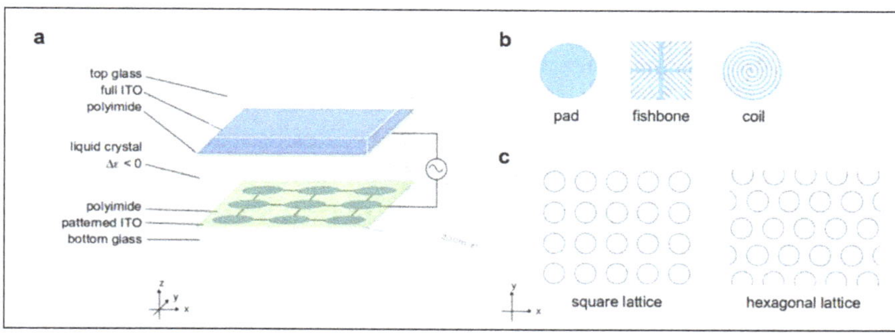

Structure of the liquid crystal cell: (a) 3D schematic diagram of the test cell. The bottom Indium-Tin-Oxide (ITO) is a pixel array. The top ITO is full.Liquid crystal is with negative dielectric anisotropy (Δε<0), and aligned vertically on the surface with the polyimide (PI). (b) Candidates of pixels. The pattern can be a pad, a fishbone, or a coil. (c) Array of pixels, arranged in square or hexagonal lattices.

A pad is the basic shape with Euler characteristic number (χ)equal to 1, which can continuously morph into a fishboneor acoil. Every pixel is a closed box with χof 2. Based on the Gauss-Bonnet and Gauss-Stein Theorem, the total topological charge of defects in a closed, homeotropic boundary is equal to χ/2. The pad, fishbone, and coil are topologically identical, so the defect generated by each of them must have the same topological charge. NLC tends to align along the strip due to the fringe field. Therefore, fishbone and coil lead to radial and circular hedgehogs, respectively. On a pad, the location of the point defect (or umbilics) is arbitrary. On fishbone and coil, the liquid crystal configuration has reflectional and rotational symmetry. Therefore, the point defects(or umbilics) can be fixed at the center of the pixel because of static balance. In this research, the splay and bend elastic constants of NLC were 15 pN, the cell gap was 3 μm, and the optimum width and spacing of the strips were chosen to be 3 μm. The size of a unit cell was 100 μm ×100 μm or50 μm ×50 μm. The pixels were connected along the axis of transnational symmetry. In square lattice, the connection between pixels were on 0°, 90°, 180° and 270°. In hexagonal lattice, the connection between pixels were on 0°, 60°, 120°, 180°, 240°, and 300°.The pixel arrays of pad, fishbone, and coil were fabricated by lithography and wet-eching. Arrays were arranged in either square or hexagonal lattices. By using the delicately designed ITO, arrays of point defects(or umbilics)were generated. The director field were analyzed under optical microscope (OM) and polarized optical microscope (POM).

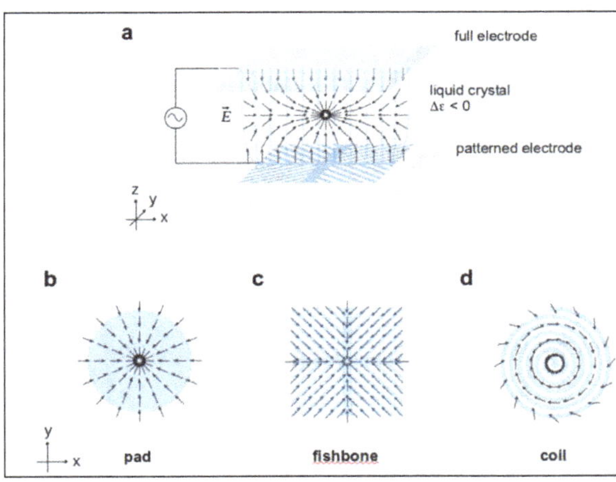

Director field in a vertically aligned liquid crystal cell.(a) 3D schematic of a pixel. The light blue, thickarrowsindicate the electric field, and the black short arrows showthe director of liquid crystal. (b), (c), (d) are top view of the pad, fishbone, and coil pixels, respectively, showing the liquid crystal configuration while electric field is applied.

Identification of Topological Defects

Pictures of bare test cells under OM is shown in figure. Every pixel is a closed homeotropic boundary with χequal to 2, so one point defect appears at each pixel, as expected by the Gauss-Stein Theorem. The dark spot is a clear sign of discontinuity in liquid crystal, because light through the defect core exhibit strong diffraction34. The destructive interference of diffracted light leads to the dark spot with texture like a hole. Light through continuous distortions, for example, umbilical escapes, shows lensing effect without a diffracting dark spot.

Additional to the point defects at the center of pixels, extra point defects appear between the pixels on the connection bars. In a square lattice, a defect in the pixel is surrounded by four extra defects. Pad, fishbone, and coil pixels result in the same structure. In hexagonal lattice, six extra defect is generated surrounding one pixel. Again, the pattern of an individual pixel does not affect the structure of defect array.

At the middle of the four (three) corners of four neighboring square (hexagonal) pixels, the director field should be a saddle deformation (saddle) as illustrated in figure. When the pixels are very close to each other, or when the electric field is strong, the saddle shrinks, and creates a director field similar to a reversed hyperbolic hedgehog. The saddle appeared as a faint, circular shade in the OM pictures. Defects and saddles can be clearly distinguished by the texture. Defect is a solid hole, and saddle is just a circular shadow. In pictures taken under POM, defect is surrounded by circular diffraction pattern, while saddles do not show diffractions. When the focus plane of OM moved from the top substrate to the bottom substrate, the image of defects evolved from concentric rings to a black dot, and then to a blurred, dark spot. This is the typical feature of Fresnel diffraction of light through a pinhole. On the other hand, when the focus plane moved from the top to the bottom, the image of saddle changed from a big faint shade to a smaller one, and then suddenly transformed into a faint, bright spot, which is barely recognized. Studies based on ray tracing in liquid crystal lens showed that a hyperbolic hedgehog act as a diverging lens. Hence, the dark shade above the saddle is a result of deflected light, and the barley recognized bright spot could be the virtual image of the diverging lens.

Point defects and saddles were arranged in square and hexagonal arrays, like atoms arranged in square and hexagonal crystal lattices. The position of the defects and saddles were fixed by the rotational symmetry in the unit cell and reflectional symmetry between the unit cells.

Director Field of the Defect Network

The director field was recognized by placing the liquid crystal cell between crossed polarizes under POM. A λ–wave plate was inserted between the cell and analyzer to add optical retardation on the 135° direction w.r.t. to the polarizer, so the directors on 45° and 135° can be clearly distinguished by the birefringence colors. Here 45° and 135° are in lime and pink, respectively. Liquid crystal on the edge of the electrode is not completely in plane, leading to blue and orange colors of shorter optical retardation.

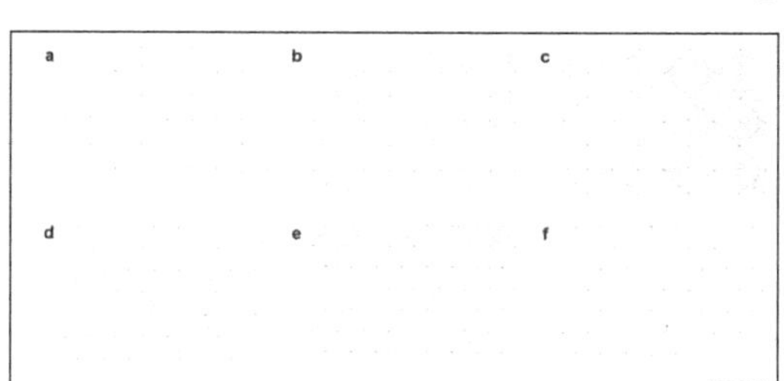

Array of point defects under optical microscope. Point defects (the black spots) in liquid crystal were generated by patterned ITO. (a) Pad pixel in square lattice. (b) Fishbone pixel in square lattice. (c) Coil pixels in square lattice. (d) Pad pixel in hexagonal lattice. (e) Fishbone pixel in hexagonal lattice. (f) Coil in hexagonal lattice.

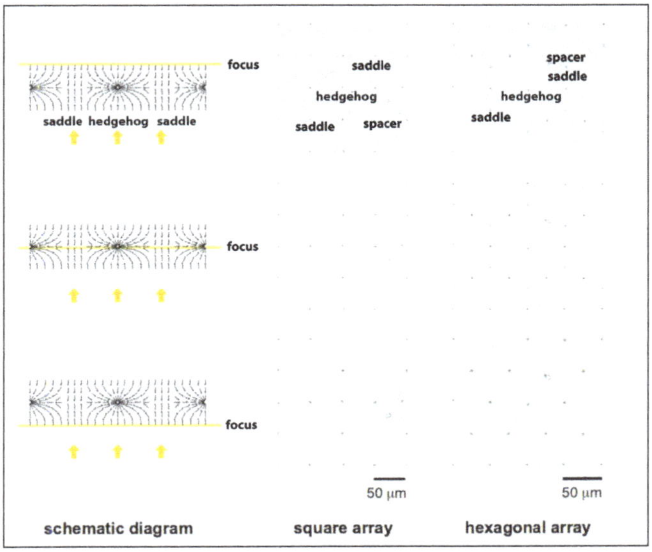

Optically sectioned images of defect array.Schematic diagram of director field is on the left. Black arrows show the directors. Yellow thick shades indicate the position of focal planes. Yellow thick arrows indicatethe direction of light. Bright field OM pictures of square and hexagonal defect array is on the right. Pictures on the top, middle, and bottom rows were taken when the focus is on the top substrate, on the defect, and on the bottom substrate, respectively. The ball spacer is the indicator for the location of the focus plane.

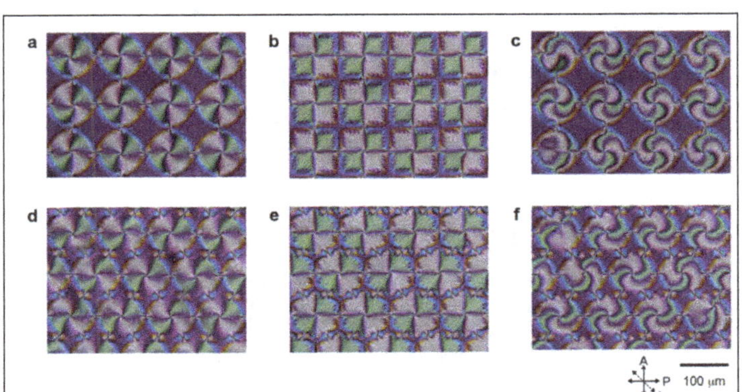

Array of point defects under polarized optical microscope. Crossed polarizers on 0° and 90°. A λ-wave plate was placed on 135°, so the directors lying in 45° and 135° show lime and pink, respectively. (a) Pad pixel in square lattice. (b) Fishbonepixel in square lattice. (c) Coil pixels in square lattice. (d) Pad pixel in hexagonal lattice. (e) Fishbone pixel in hexagonal lattice. (f) Coil in hexagonal lattice.

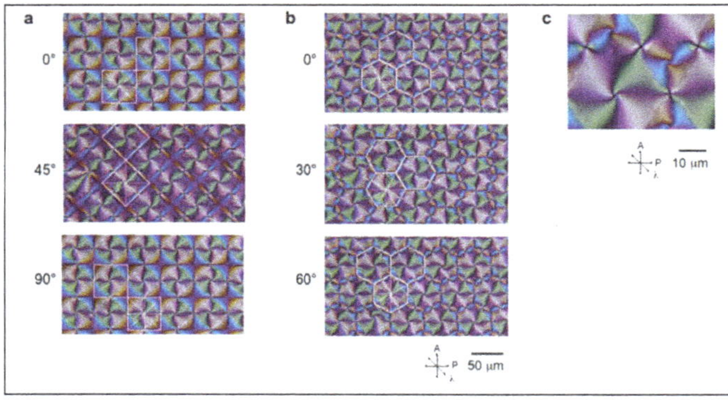

Rotation of defect array under polarized microscope. Crossed polarizers on 0° and 90°. λ-wave plate on 135°. (a)Square array lying in 0°, 45° and 90°. (b)Hexagonal array lying in 0°, 30° and 60°.White square and hexagonal box indicates the domain boundary of a unit cell. Dashed, white lines show the simplified director field. (c) Diffraction of light through the point defects.

Topological charge of the defects was identified by Shlieren textures under POM. The topological charge is the number of brushes divided by 4. All the defects had four dark brushes, meaning that the topological charge was either +1 or -1. Sign of defect was examined by rotating the polarizers. All the defects at the center of the pixel rotated at the same direction with the polarizes, but the defects on the connecting bars of pixels rotated in an opposite direction. Therefore, the defect at the pixel was a radial (or spiral) hedgehog, surrounded by four (six) hyperbolic hedgehogs in square (hexagonal) lattices. In 2D space, topological charges of radial and hyperbolic hedgehog are +1 and -1, respectively. The network of hyperbolic hedgehogs is like the scaffold of the 2D crystal. The structure of the scaffold is determined by the shape of the lattice (square or hexagonal), unaffected by the radial (or spiral) hedgehog at the center of the pixel.

The rotational symmetry in the director field was identified by rotating the test cell while the polarizer, analyzer, and wave plate were fixed. The domain boundaries of a unit cell were recognized by the rotation, too. When the boundary was parallel to the polarizer or the analyzer, it shows a thin, dark line. Every 90°of rotation of a square array gave the same optical image, meaning that the director field had 4-fold rotational symmetry. The axis of mirror symmetry of a hyperbolic hedgehog became the domain boundary between the unit cells. Each hyperbolic hedgehog was shared by two unit cells (white solid lines. The saddle was shared by four neighboring unit cells. On the other hand, the hexagonal array had6-fold rotational symmetry, since every 60°of rotation gave the same optical image. Every side of the hexagon (walls in 3D) has a half hyperbolic hedgehog embedded on the surface. Note that the saddle here is shared by three neighboring corners of three hexagons. The hyperbolic hedgehog between two unit cells splits the two neighboring unit cells.

Referring to the colors of the POM images, the director field of the topological defect network was estimated. A pad generates the radial hedgehog. A fishbone electrode simply emphasizes the diagonal components of the radial hedgehog. In a coil pixel, the directors circle around the defect, generating a spiral hedgehog. Circular hedgehogs were not observed. The director field inside the pixel are successfully controlled by the strips of the electrodes, but not affected by the lattice type (square or hexagonal).

Figure shows the topological defect array in a wide range. The scale of each picture is 1.7 mm ×1.2 mm. The square and fishbone pixels could form a regular, periodic, stable topological defect network. The size of defect network could be larger than 1 cm². In contrast, the spiral hedgehogs generated by coil pixels could not be arranged in a regular array. Most of the spiral hedgehogs deviated from the center of the pixel, or transformed into radial hedgehogs. Therefore, the distribution of director field was irregular. The spiral hedgehogs could easily drift away from the designed position, merged with the hyperbolic hedgehog nearby, and then the defects disappeared. The defect network composed of spiral hedgehogs was instable. To summarize, arrays of point defects were successfully generated in vertically aligned liquid crystal cell when electric field was applied. All defects were points. Lines and disclinations had not been observed. The radial hedgehogs were positioned by design, and the hyperbolic hedgehogs where generated on the domain boundary spontaneously. Defects were arranged in square or hexagonal lattices like square or hexagonal crystals.

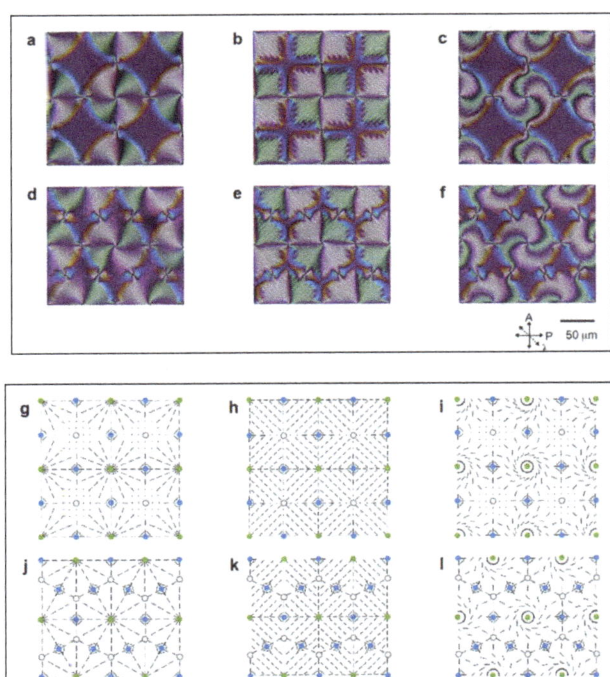

Director field of defect arrays: (a)-(f) Pictures of square and hexagonal arrays under polarized microscope. Crossed polarizers on 0° and 90°. λ-wave plate on 135°.(g)–(l) Corresponding director fields.(a) (g) Pad pixel in square lattice. (b) (h) Fishbone pixel in square lattice. (c) (i) Coil pixels in square lattice. (d) (j) Pad pixel in hexagonal lattice. (e) (k) Fishbone pixel in hexagonal lattice. (f) (l) Coil in hexagonal lattice. Green, blue, and white dots indicate the position of radial hedgehog, hyperbolic hedgehog, and saddle deformations, respectively. The black short lines show the director of liquid crystal.

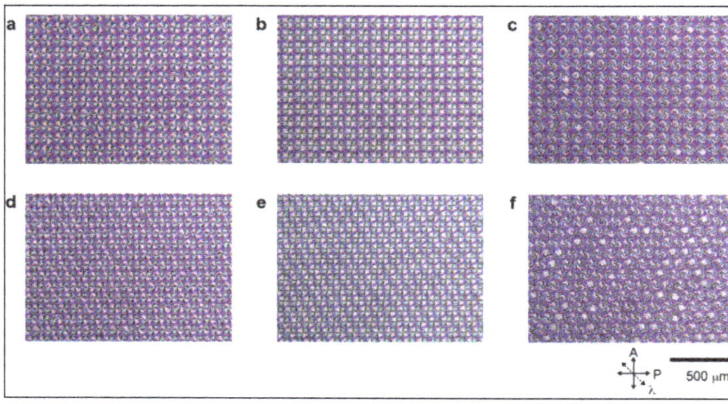

Large range of defect arrays under polarized optical microscope. Crossed polarizers on 0° and 90°.λ-wave plate on 135°. (a) Pad pixel in square lattice. (b) Fishbone pixel in square lattice. (c) Coil pixels in square lattice. (d) Pad pixel in hexagonal lattice. (e) Fishbone pixel in hexagonal lattice. (f) Coil in hexagonal lattice.

Analysis

The radial (or spiral) hedgehogs, hyperbolic hedgehogs, and saddles were successfully arranged in two dimensional (2D) square and hexagonal lattices. The arrays of topological defects were stable, and span over a large area. Though the information on the 3rd dimension (the z-axis) is limited under POM, the author does not want to be limited in two dimensions. The three dimensional (3D) unit cells can be deduced based on principles of symmetry and continuity of director field. The director field on layer II in figure was clearly identified in POM images. Layer I and III are above and beneath Layer II, respectively. The defects on layer I and III can be derived according to the following principles:

- A saddle iscreated at the middle of two out-going radial hedgehogs.
- A hyperbolic hedgehog is created at the middle of two saddles.
- The saddles and hyperbolic hedgehogs generate a boundary of which all the directors are pointing into the unit cell, resulting in the in-going radial hedgehog.

The topological defects on layer I and III can be derived. Therefore the3D cuboid and hexagonal cylinder unit cells are postulated, respectively. The director field and topological charges of the 3D unit cells were analyzed to find the hidden conservation laws.

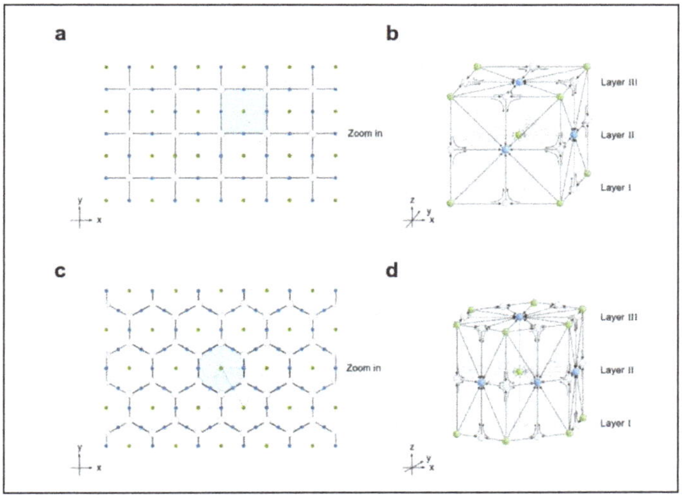

Crystal of topological defects and unit cells. (a) 2D square crystal (b) 3D cuboid unit cell (c) 2Dhexagonal crystal (d) 3D hexagonal cylinder unit cell. The area in light blue are the unit cell of the crystal. Black lines show the domain boundary. Green, blue, and white dots indicate the position of radial hedgehog, hyperbolic hedgehog, and saddle deformations, respectively. The unit cell of a hexagonal crystal can be a hexagon or two anti-parallel triangles (marked by the red dashed lines). The black arrows show the director field.

Note that the axis of rotational symmetry (C_∞) of hyperbolic hedgehogs are normal to the surface of unit cells. The director field of a saddle is very similar to that of a hyperbolic hedgehog, but all the directorsare reversed, which leads to a reversed hyperbolic hedgehog. The C_∞ of saddles is along the edge of unit cells. The liquid crystal molecules are considered as vectors here, though they have the head-to-tail symmetry. Due to the anchoring on the surface, the head-to-tail symmetry is broken. The end fixed on the surface is defined to be the tail. The other end that can move freely with respect to the applied electric field is defined as the head. The vectors are indicated by arrows.

Conservation of Director Flux

The director field created by topological defects seems very similar to the electric field created by electric charges. Though the electric field is a graphical representation of force and it does not move (but the force drives the charge moving), the electric field ($\vec{E}$) and charges(q) are still considered as flux and its sources (or sinks) in Electromagnetism, respectively. The relation between the flux and source is indicated in Gauss Law:

$$\oiint \vec{E} \cdot d\vec{A} = \frac{q}{\varepsilon}$$

where $d\vec{A}$ is the unit area indicated by a normal vector, ε is dielectric constant, q/ε represents the strength of the source (or the sink) in terms of electric charge (q).

The relation between the director flux and the topological defects are postulated in the following equation:

$$\oiint \vec{n} \cdot d\vec{A} = \text{strength of source or sink}$$

where the strength of the source (or sink) is proportional to the flux of directors from (or to) the topological defects.

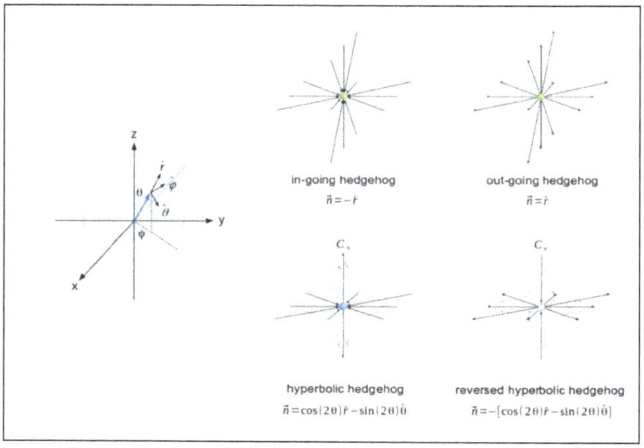

Director field around point defects in details. Black and grey arrows show the director field. C_∞ marks the axis of rotational symmetry. The planes are the mirror of reflection symmetry. Spherical coordinate system on the left hand side. Mathematical representations of director field are under each schematic diagram.

Flux in Cuboid Crystal

The flux of directors from (or into) a topological defect is calculated below. The director field of radial hedgehogs are described by the following expressions in spherical coordinates:

$$\text{Out} - \text{going radial hedgehog}: \vec{n} = \hat{r}$$

$$\text{In} - \text{going radial hedgehog}: \vec{n} = -\hat{r}$$

Where $\vec{n}$ is the director and $\hat{r}$ is the unit vector on the radial direction. The flux of directors from a radial hedgehog (Φ_{hr+}) is the integration of $\vec{n} \cdot d\vec{A}$ over a closed surface that wraps the hedgehog up:

$$\Phi_{hr+} = \oiint \vec{n} \cdot d\vec{A} = \oiint \hat{r} \cdot d\vec{A} = \oiint_{2\pi} \hat{r} \cdot \left(rd\theta\hat{\theta} \times r\sin\theta \, d\phi\hat{\phi} \right)$$

$$= r^2 \int_0^{\pi} \sin\theta \, d\theta \int_0^{2\pi} d\phi = 4\pi r^2$$

The flux of directors flowing into a radial hedgehog (Φ_{hr-}) is:

$$\Phi_{hr-} = \oiint \vec{n} \cdot d\vec{A} = \oiint -\hat{r} \cdot d\vec{A} = -4\pi r^2$$

Because $4\pi r^2$ is the surface area of the sphere, which scales with the radius r, the strength of radial hedgehog as a source (or sink) can be defined by the flux per unit area, namely, the "director flux charge". The director flux charge of an out-going radial hedgehog (Q_{hr+}) and an in-going radial hedgehog (Q_{hr-}) are given by the calculation below:

$$Q_{hr+} = \frac{\Phi_{hr+}}{4\pi r^2} = +1$$

$$Q_{hr-} = \frac{\Phi_{hr-}}{-4\pi r^2} = -1$$

The director field of a hyperbolic hedgehog is given by the equation:

$$\text{hyperbolic hedgehog} : \vec{n} = \left[\cos(2\theta)\hat{r} - \sin(2\theta)\hat{\theta}\right]$$

Here the hyperbolic hedgehog has the rotational symmetry about the z-axis, and θ is the angle between $\vec{n}$ and the axis of rotational symmetry. The reversed hyperbolic hedgehog is obtained by reversing the directors of a hyperbolic hedgehog:

$$\text{Reversed hyperbolic hedgehog} : \vec{n} = -\left[\cos(2\theta)\hat{r} - \sin(2\theta)\hat{\theta}\right]$$

Saddle deformation and reversed hyperbolic hedgehog have the same director field, though saddle does not have a defect core. The flux of directors from a hyperbolic hedgehog (Φ_{hh+}) is:

$$\begin{aligned}\Phi_{hh+} &= \oiint \vec{n} \cdot d\vec{A} \\ &= \oiint \left[\cos(2\theta)\hat{r} - \sin(2\theta)\hat{\theta}\right] \cdot \left(r d\theta\hat{\theta} \times r \sin\theta \, d\phi\hat{\phi}\right) \\ &= r^2 \int_0^{\pi} \cos(2\theta) \sin\theta \, d\theta \int_0^{2\pi} d\phi = r^2 \\ &= -\frac{4}{3}\pi r^2\end{aligned}$$

The flux from a reversed hyperbolic hedgehog (Φ_{hh-}) is:

$$\Phi_{hh-} = \oiint \vec{n} \cdot d\vec{A}$$

$$= \oiint -\left[\cos(2\theta)\hat{r} - \sin(2\theta)\hat{\theta}\right] \cdot \left(rd\theta\hat{\theta} \times r\sin\theta\, d\phi\hat{\phi}\right) = \frac{4}{3}\pi r^2$$

Then the director charge of a hyperbolic hedgehog (Q_{hh+})and a reversed hyperbolic hedgehog (Q_{hh-}) are given as below:

$$Q_{hh+} = \frac{\Phi_{hh+}}{4\pi r^2} = -\frac{1}{3}$$

$$Q_{hh-} = \frac{\Phi_{hh-}}{4\pi r^2} = \frac{1}{3}$$

Directors are from the out-going radial hedgehogs on the 8 corners of the unit cell. They turn the direction at the hyperbolics and saddles, and finally gather at the in-going radial hedgehog at the center of the cell. The structure of layer I and III are the same. The directors on the walls were in plane, but they switch to the direction normal to the wall via the saddles and hyperbolic hedgehogs on Layer II. Finally, all the directors "flow" into the center of the unit cell, and the in-going radial hedgehog is generated. On layer I and III, an out-going radial hedgehog is shared by 8neighboring unit cells, one saddle is shared by 4 neighboring unit cells, and the hyperbolic hedgehog at the center of a wall is shared by 2 neighboring unit cells. On layer II, an in-going hedgehog is entire lyin the unit cell, a saddle on the corner is shared by 4neighboring cells, and a hyperbolic hedgehog on the side is shared by 2 neighboring unit cells. The flux charges on each layer:

Centre Corner Sides

$$\text{Layer I}: 1\times\left[\frac{1}{2}\text{hyperbolic hedgehog}\right] + 4\times\left[\frac{1}{8}\text{out-going radial hedgehog}\right] + 4\times\left[\frac{1}{4}\text{saddle}\right]$$

$$\text{Layer II}: 1\times\left[1\,\text{in-going radial hedgehog}\right] + 4\times\left[\frac{1}{4}\text{saddle}\right] + 4\times\left[\frac{1}{2}\text{hyperbolic hedgehog}\right]$$

$$\text{Layer III}: 1\times\left[\frac{1}{2}\text{hyperbolic hedgehog}\right] + 4\times\left[\frac{1}{8}\text{out-going radial hedgehog}\right] + 4\times\left[\frac{1}{4}\text{saddle}\right]$$

Therefore,

Centre Corner Sides

$$\text{Layer I}: 1\times\left[\frac{1}{2}Q_{hh+}\right] + 4\times\left[\frac{1}{8}Q_{hr+}\right] + 4\times\left[\frac{1}{4}Q_{hh-}\right]$$

$$\text{Layer II}: 1\times\left[1Q_{hr-}\right] + 4\times\left[\frac{1}{4}Q_{hh-}\right] + 4\times\left[\frac{1}{2}Q_{hh+}\right]$$

$$\text{Layer III}: 1\times\left[\frac{1}{2}Q_{hh+}\right] + 4\times\left[\frac{1}{8}Q_{hr+}\right] + 4\times\left[\frac{1}{4}Q_{hh-}\right]$$

$$\begin{array}{llll} & \text{Centre} & \text{Corner} & \text{Sides} \\ \text{Layer I}: Q_I = & 1\times\left[\frac{1}{2}\left(-\frac{1}{3}\right)\right] & +4\times\left[\frac{1}{8}(+1)\right] & +4\times\left[\frac{1}{4}\left(1+\frac{1}{3}\right)\right] = \frac{1}{6}+\frac{1}{2}+\frac{1}{3} \\ \text{Layer II}: Q_{II} = & 1\times\left[1-(-1)\right] & +4\times\left[\frac{1}{4}\left(+\frac{1}{3}\right)\right] & +4\times\left[\frac{1}{2}\left(-\frac{1}{3}\right)\right] = 1+\frac{1}{3}-\frac{2}{3} \\ \text{Layer III}: Q_{III} = & 1\times\left[\frac{1}{2}\left(-\frac{1}{3}\right)\right] & +4\times\left[\frac{1}{8}(+1)\right] & +4\times\left[\frac{1}{4}\left(+\frac{1}{3}\right)\right] = \frac{1}{6}+\frac{1}{2}+\frac{1}{3} \end{array}$$

The total "flux charge" in a lattice is:

$$Q_I + Q_{II} + Q_{III} = 0$$

which reveals that the total director flux in a cuboid unit cell is 0, indicating that the directors pointing into the unit cell must equal to the directors gathered at the center defect. The director flux is conserved in a stable cuboid unit cell. Since crystal is a periodic repetition of unit cells, the flux is conserved in the whole defect crystal, too.

The cuboid unit cell with a radial hedgehog at the center, created by the pad and fishbone electrode, which leads to stable and robust crystal of defects, is the positive case for flux conservation. On the other hand, the circular hedgehog at a cuboid unit cell, created by coil electrode, is instable in the defect array, which is the negative example for flux conservation. The center defect turns into a spiral hedgehog (a mixture of radial and circular components), or it simply shows up as a radial hedgehog of which the location is out of control.

The director field of a circular hedgehog is,

$$\vec{n} = \hat{\phi}$$

and the flux of a circular hedgehog (Φ_{hc}) is

$$\Phi_{hc} = \oint\!\!\oint \vec{n} \cdot d\vec{A} = \oint\!\!\oint \hat{\phi} \cdot \left(r d\theta\hat{\theta} \times r\sin\theta\, d\phi\hat{\phi} \right) = 0$$

The flux of directors is a closed loop in the circular defect, so the flux is zero. Thus the flux charge (Q_{hc}) is 0, too:

$$Q_{hc} = 0$$

In this case, the total director flux in a unit cell is not zero:

$$Q_I + Q_{II} + Q_{III} = +1 \neq 0$$

Indicating that the directors pointing into the unit cell is more than the directors absorbed by the center defect. The flux from the source is not consistent with the flux to the sink, and therefore the crystal of defect is not stable. The directors from the source rush into unexpected directions to release the flux to the sink. Hence the crystal of circular hedgehogs collapse in periodic homeotropic boundaries.

Flux in Hexagonal Cylinder Crystal

In a hexagonal cylinder unit cell, the directors are from twelve out-going radial hedgehogs on layer I and layer III. The planar directors on the walls switch into a vertical alignment via the hyperbolic hedgehog sand saddles. Finally, all the directors point into the center of the unit cell, generating an in-going radial hedgehog. Actually, the unit cells in a hexagonal crystal must be triangles arranged in anti-parallel manner. Each corner of the triangular unit cell is 1/6 of an in-going radial hedgehog. Therefore, a saddle contributes only 3×(1/6) of director flux on Layer II. Here, we keep using the hexagonal cylinder unit cell to have consistent analysis on the data and in the models. Distribution of hyperbolic hedgehogs and saddles on the walls are the same as the cases in cuboid unit cells. The three layers of defect network are indicated in figure and summarized the following equations:

$$\begin{array}{llll} & \text{Centre} & \text{Corner} & \text{Sides} \\ \text{Layer I}: & 1\times\left[\frac{1}{2}\text{hyperbolic hedgehog}\right] & +6\times\left[\frac{1}{6}\cdot\frac{1}{2}\text{out-going radial hedgehog}\right] & +6\times\left[\frac{1}{4}\text{saddle}\right] \\ \text{Layer II}: & 1\times\left[1\,\text{in-going radial hedgehog}\right] & +6\times\left[\frac{1}{3}\cdot\frac{1}{2}\text{saddle}\right] & +6\times\left[\frac{1}{2}\text{hyperbolic hedgehog}\right] \\ \text{Layer III}: & 1\times\left[\frac{1}{2}\text{hyperbolic hedgehog}\right] & +6\times\left[\frac{1}{6}\cdot\frac{1}{2}\text{out-going radial hedgehog}\right] & +6\left[\frac{1}{4}\text{saddle}\right] \end{array}$$

Then the flux chargeson each layer are:

$$\begin{array}{llll} & \text{Centre} & \text{Corner} & \text{Sides} \\ \text{Layer I}: & 1\times\left[\frac{1}{2}Q_{hh+}\right] & +6\times\left[\frac{1}{12}Q_{hr+}\right] & +6\times\left[\frac{1}{4}Q_{hh-}\right] \\ \text{Layer II}: & 1\times\left[1Q_{hr-}\right] & +6\times\left[\frac{1}{6}Q_{hh-}\right] & +6\times\left[\frac{1}{2}Q_{hrr+}\right] \\ \text{Layer III}: & 1\times\left[\frac{1}{2}Q_{hh+}\right] & +6\times\left[\frac{1}{12}Q_{hr+}\right] & +6\times\left[\frac{1}{4}Q_{hh-}\right] \end{array}$$

$$\begin{array}{lllll} & \text{Centre} & \text{Corner} & \text{Sides} & \\ \text{Layer I}: Q_{I} = & 1\times\left[\frac{1}{2}\left(-\frac{1}{3}\right)\right] & +6\times\left[\frac{1}{12}(+1)\right] & +6\times\left[\frac{1}{4}\left(+\frac{1}{3}\right)\right] & =-\frac{1}{6}+\frac{1}{2}+\frac{1}{2} \\ \text{Layer II}: Q_{II} = & 1\times\left[1(-1)\right] & +6\times\left[\frac{1}{6}\left(+\frac{1}{3}\right)\right] & +6\times\left[\frac{1}{2}\left(-\frac{1}{3}\right)\right] & =-1+\frac{1}{3}-1 \\ \text{Layer III}: Q_{III} = & 1\times\left[\frac{1}{2}\left(-\frac{1}{3}\right)\right] & +6\times\left[\frac{1}{12}(+1)\right] & +6\times\left[\frac{1}{4}\left(+\frac{1}{3}\right)\right] & =-\frac{1}{6}+\frac{1}{2}+\frac{1}{2} \end{array}$$

The total "director charge" in a lattice is:

$$Q_I + Q_{II} + Q_{III} = 0$$

The total flux is 0 again. The flux from the source is equal to the flux to the sink. The director flux is conserved in hexagonal unit cell, too. However, there is more radial component in a hexagonal unit cell than in a cuboid one, since hexagonal unit cells have more saddles and hyperbolic hedgehogs on the corners and edges. Therefore, radial and spiral hedgehogs are stabilized, and circular hedgehogs are not favored in hexagonal crystals.

Conservation of Topological Charges

The summation of topological charges in confined liquid crystal is determined by the Euler characteristic (χ) of the confinement. To have a clear logic flow in the following derivation, topological charges are classified into surface charges (s) and bulk charges (m_b). For a closed confinement with planar surface the summation of surface charges (s_i) is:

$$\sum_i s_i = \chi$$

where i is the index. The equation was derived from Poincaré-Hopf theorem. For homeotropic surfaces the total bulk charges in the confinement is:

$$\sum_i m_{b,k} = \frac{\chi}{2}$$

where k is the index. The relation is from the Gauss-Stein Theorem for characteristic of surfaces and Gauss-Bonnet Theorem for singularities in vector field. The sum of surface charge on a homeotropic confinement must be 0. The Euler characteristic is topologically invariant. Though the number of charges can change due to phase transition, Fréedericksz transition, or variation of geometries, the summation of topological charges must conserve. χ of a closed sphere is 2. A cuboid or hexagonal cylinder surface is topologically equivalent to a sphere, so the χ of our unit cell is always 2. A single unit cell with homeotropic surface obeys the equation, $\sum_k m_{b,k} = \chi/2$. However, additional defects are created on the boundary between unit cells in the lattice. If topological charge is conserved in a closed confinement, the total sand m_b must be 0 and 1, respectively, for both cuboid and hexagonal cylinder unit cells. Additionally, the m_b inside the confinement and the summation of m_b embedded on the boundary must equal to each other, because the degree of continuous mapping must conserve, too. The bulk charge is the wrapping number of closed surfaces surrounding a defect, which can be calculated by the following equation:

$$m_b = \frac{1}{4\pi} \oiint \left[\vec{n} \cdot \left(\frac{\partial \vec{n}}{\partial \theta} \times \frac{\partial \vec{n}}{\partial \phi} \right) \right] d\theta d\phi$$

The projection of 3D director field around a bulk defect onto the surface of its confinement is the 2D surface defect, and its strength is the winding number of closed loops surrounding the defect core:

$$s = \frac{1}{2\pi}\oint\left(\vec{n}\cdot\frac{\partial \vec{n}}{\partial \phi}\right)\cdot d\vec{l}$$

The m_b and s of radial hedgehogs are always +1. The m_b of regular and reversed hyperbolic hedgehogs are +1 and -1, respectively. The s of hyperbolic hedgehogs depends on the surface for projection. When the C_∞ axis of a hyperbolic hedgehog is the z-axis, its s is +1 on the x-y plane, but its projection on y-z plane results in s equal to -1. The s of a reversed hyperbolic hedgehog is obtained in the same manner. The positions of the topological defects are illustrated in figure cuboid and hexagonal cylinder unit cells, respectively. The m_b and s are marked next to the defects. In a cuboid unit cell, the m_b of radial hedgehog inside the cell is +1. The m_b of defects embedded on the surfaces are summarized in the equations below:

Centre Corner Sides

$$\textbf{Layer I}: m_{b,I} = 1\times\left[\frac{1}{2}(+2)\right]+4\times\left[\frac{1}{8}(+1)\right]+4\times\left[\frac{1}{4}(-1)\right]=+\frac{1}{2}+\frac{1}{2}-1$$

$$\textbf{Layer II}: m_{b.II} = 0\times\left[0(+0)\right]+4\times\left[\frac{1}{4}(-4)\right]+4\times\left[\frac{1}{2}(+1)\right]=+0-1+2$$

$$\textbf{Layer III}: m_{b,III} = \times\left[\frac{1}{2}(+1)\right]+4\times\left[\frac{1}{8}(+1)\right]+4\times\left[\frac{1}{4}(-1)\right]=+\frac{1}{2}+\frac{1}{2}-1$$

The total m_b of defects embedded on the surface is then,

$$m_{b,I}+m_{b,II}+m_{b,III}=+1$$

The bulk topological charge obeys the Gauss-Stein Theorem for homeotropic boundary. Then the m_b of center bulk defect is equivalent to the total m_b of surface defects because the degree of mapping conserves.

The surface charge, s, of each wall of the unit cell is summarized in the following equation:

Centre Corner Sides

$$\textbf{Ceiling}: S_{ceiling} = 1\times[+1]+4\times\left[+\frac{1}{4}\right]+4\times\left[-\frac{1}{2}\right]=+1+1-2$$

$$\textbf{Wall}: S_{wall} = 1\times[+1]+4\times\left[+\frac{1}{4}\right]+4\times\left[-\frac{1}{2}\right]=+1+1-2$$

$$\textbf{Floor}: S_{floor} = 1\times[+1]+4\times\left[+\frac{1}{4}\right]+4\times\left[-\frac{1}{2}\right]=+1+1-2$$

The total s of defects projected on the boundary is then:

$$s_{ceiling} + 4 \cdot s_{wall} + s_{floor} = 0$$

which is identical to the total s of a homeotropic surface.

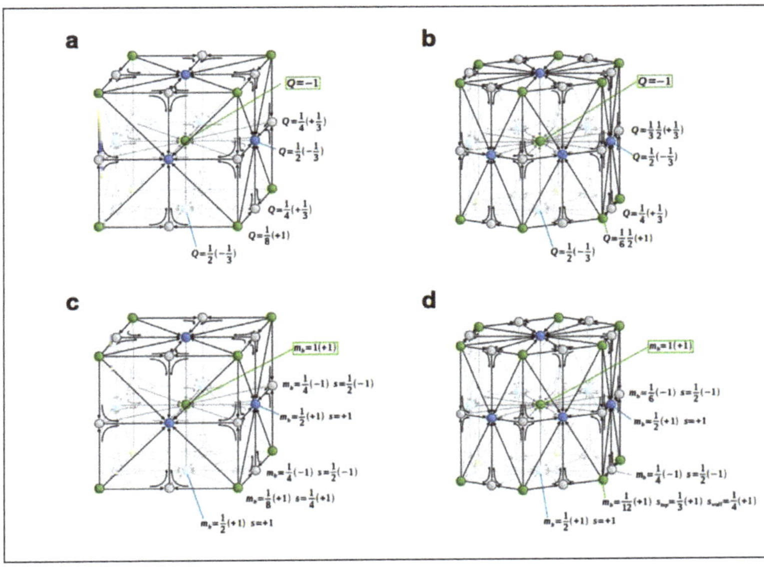

Charges of unit cells. (a)Flux charge of director flux in a cuboid unit cell. (b) Flux charge in a hexagonal unit cell. (c)Topological charge in a cuboid unit cell. (d) Topological charge in a hexagonal unit cell. Q is flux charge. m_b and s are bulk and surface topological charge, respectively. Green, blue, and white dots indicate the position of radial hedgehog, hyperbolic hedgehog, and saddle deformations, respectively. The black arrows show the director field.

The same analysis is performed for the hexagonal crystal. In a hexagonal cylinder unit cell, the m_b inthe bulk is +1, too. The m_b of defects embedded on the surfaces are summarized in the following equations:

Centre **Corner** **Sides**

$$\textbf{Layer I}: m_{b,I} = 1\times\left[\frac{1}{2}(+1)\right] + 6\times\left[\frac{1}{12}(+1)\right] + 6\times\left[\frac{1}{4}(-1)\right] = +\frac{1}{2} + \frac{1}{2} - \frac{3}{2}$$

$$\textbf{Layer II}: m_{b.II} = 0\times\left[0(+0)\right] + 6\times\left[\frac{1}{6}(-1)\right] + 6\times\left[\frac{1}{2}(+1)\right] = +0 - 1 + 3$$

$$\textbf{Layer III}: m_{b,III} = \times\left[\frac{1}{2}(+1)\right] + 6\times\left[\frac{1}{12}(+1)\right] + 6\times\left[\frac{1}{4}(-1)\right] = +\frac{1}{2} + \frac{1}{2} - \frac{3}{2}$$

$$m_{b,I} + m_{b,II} + m_{b,III} = +1$$

For the surface defects:

Centre Corner Sides

$$\textbf{Ceiling}: S_{ceiling} = 1\times[+1]+6\times\left[+\frac{1}{3}\right]+6\times\left[-\frac{1}{2}\right] = +1+2-3$$

$$\textbf{Wall}: \quad S_{wall} = 1\times[+1]+4\times\left[+\frac{1}{4}\right]+4\times\left[-\frac{1}{2}\right] = +1+1-2$$

$$\textbf{Floor}: S_{floor} = 1\times[+1]+6\times\left[+\frac{1}{4}\right]+6\times\left[-\frac{1}{2}\right] = +1+2-3$$

$$s_{ceiling} + 6\cdot s_{wall} + s_{floor} = 0$$

In both cuboid and hexagonal cylinder unit cells:

$$\sum_{bulk} m_b = \sum_{surface} m_b = \frac{\chi}{2}$$

$$\sum_{surface} s = 0$$

The bulk topological charge is determined by the Euler characteristic of the confinement. If the shapes of confinements are topologically invariant, the total m_b conserves. The surface charge is 0 for homeotropic, closed surface. It is independent of the shape of confinement, and conserve under continuous morphing, too.

Note that the radial and circular hedgehog has the identical m_b and s, which are equal to1. If a radial hedgehog were replaced by a circular one at the center of unit cells, the total m_b and s would be unchanged. However, the experimental results show that circular hedgehog is instablein crystal of defects. The difference between radial and circular hedgehogs is well explained in terms of conservation flux: The total flux is 0 if the bulk defect is a radial hedgehog, while the flux from the source and the flux to the sink cannot be equal if the bulk defect is a circular hedgehog. Radial and circular hedgehogs are identical in the analysis of topological charges, so the conservation of topological charges is notcapableto predict the stability of defect network in this case.

For crystal of defects generated in homeotropic, period confinement, conservation laws are summarized in the following equations:

$$\sum_{bulk} Q = 0$$

$$\sum_{bulk} m_b = \frac{\chi}{2}$$

$$\sum_{surface} s = 0$$

The flux (represented by charge, Q), bulk topological charge (m_b), and surface topological charge (s) conserve under continuous morphing of the unit cells, and must be invariant under deformations.

Physical Meaning of Flux Charge Conservation

Flux charge represents the divergence of director field round the defect. Divergence of the director field is a measurement for splay deformation, which is dominant in liquid crystal confined in homeotropic surfaces. The elastic potential energy is proportional to the square of the divergence, and the gradient of the divergence is the force between defects in a defect crystal. The self-organizing behavior of particles in liquid crystal is driven by this kind of long-range force. The conservation of flux charge implies that the long-range force is balanced and the potential energy is minimized. The crystal of topological defects mist be a ground state of energy and thus stabilized.

Physical Meaning of Topological Charge Conservation

The conservation of topological charge is determined and protected by the topological structure of the confinement. It usually applies to soft, self-retained, self-healing structure because the topological charges are invariant under continuous distortion. Topological charge conservation also implies that the sufficient combination of defects in a stable defect crystal should be quantized.

Table: Flux, bulk, and surface charges of hedgehogs.

	Out-going radial hedgehog(prefix: hr+)	In-going radial hedgehog (prefix: hr-)	Hyperbolic hedgehog(prefix: hh+)	Reversed hyperbolic hedgehog(prefix: hh-)
Flux charge (Q)	+1	+1	$-\frac{1}{3}$	$+\frac{1}{3}$
Bulk topological charge (m_b)	+1	+1	+1	-1
Surface topological charge (s) on x-y plane	+1	+1	+1	+1
Surface topological charge (s) on y-z plane	+1	+1	-1	-1

References

- Discrete-lattice-theory-for-Frenkel-defect-aggregation-in-irradiated-ionic-solids-233996523: researchgate.net, Retrieved 16 April, 2020
- Cruz, M.; Turok, N.; Vielva, P.; Martínez-González, E.; Hobson, M. (2007). "A Cosmic Microwave Background Feature Consistent with a Cosmic Texture". Science. 318 (5856): 1612–1614. arXiv:0710.5737. Bibcode:2007Sci...318.1612C. doi:10.1126/science.1148694. PMID 17962521

- Surface-defects-in-crystals-3-types-metallurgy, defects-metallurgy, metallurgy-43279: engineeringe-notes.com, Retrieved 07 August, 2020
- Forsblom, M.; Grimvall, G. (2005). “Homogeneous melting of superheated crystals: Molecular dynamics simulations”. Phys. Rev. B. 72 (5): 054107. Bibcode:2005PhRvB..72e4107F. doi:10.1103/PhysRevB.72.054107
- Schottky-defect, chemistry: byjus.com, Retrieved 19 April, 2020

6

Diverse Aspects of Crystallography

Diverse significant aspects are studied under the domain of crystallography which include crystallization. This chapter has been carefully written to provide an easy understanding of the important aspects of crystallography.

Crystal defect is imperfection in the regular geometrical arrangement of the atoms in a crystalline solid. These imperfections result from deformation of the solid, rapid cooling from high temperature, or high-energy radiation (X-rays or neutrons) striking the solid. Located at single points, along lines, or on whole surfaces in the solid, these defects influence its mechanical, electrical, and optical behaviour.

Crystallization

Crystallization is the solidification of atoms or molecules into a highly structured form called a crystal. Usually, this refers to the slow precipitation of crystals from a solution of a substance. However, crystals can form from a pure melt or directly from deposition from the gas phase. Crystallization can also refer to the solid-liquid separation and purification technique in which mass transfer occurs from the liquid solution to a pure solid crystalline phase.

Although crystallization may occur during precipitation, the two terms are not interchangeable. Precipitation simply refers to the formation of an insoluble (solid) from a chemical reaction. A precipitate may be amorphous or crystalline.

The Process of Crystallization

Two events must occur for crystallization to occur. First, atoms or molecules cluster together on the microscopic scale in a process called nucleation. Next, if the clusters become stable and sufficiently large, crystal growth may occur.

Atoms and compounds can generally form more than one crystal structure (polymorphism). The arrangement of particles is determined during the nucleation stage of crystallization. This may be influenced by multiple factors, including temperature, the concentration of the particles, pressure, and the purity of the material.

In a solution in the crystal growth phase, an equilibrium is established in which solute particles dissolve back into the solution and precipitate as a solid. If the solution is supersaturated, this drives crystallization because the solvent cannot support continued dissolving. Sometimes having a supersaturated solution is insufficient to induce crystallization. It may be necessary to provide a seed crystal or a rough surface to start nucleation and growth.

Examples of Crystallization

A material may crystallize either naturally or artificially and either quickly or over geological timescales. Examples of natural crystallization include:

- Snowflake formation.
- Crystallization of honey in a jar.
- Stalactite and stalagmite formation.
- Gemstone crystal deposition.

Examples of artificial crystallization include:

- Growing sugar crystals in a jar.
- Production of synthetic gemstones.

Crystallization Methods

There are many methods used to crystallize a substance. To a large degree, these depend on whether the starting material is an ionic compound (e.g., salt), covalent compound (e.g., sugar or menthol), or a metal (e.g., silver or steel). Ways of growing crystals include:

- Cooling a solution or melt.

- Evaporating a solvent.
- Adding a second solvent to reduce the solubility of the solute.
- Sublimation.
- Solvent layering.
- Adding a cation or anion.

The most common crystallization process is to dissolve the solute in a solvent in which it is at least partially soluble. Often the temperature of the solution is increased to increase solubility so the maximum amount of solute goes into solution. Next, the warm or hot mixture is filtered to remove undissolved material or impurities. The remaining solution (the filtrate) is allowed to slowly cool to induce crystallization. The crystals may be removed from the solution and allowed to dry or may be washed using a solvent in which they are insoluble. If the process is repeated to increase the purity of the sample, it is called recrystallization.

The rate of cooling of the solution and the amount of evaporation of the solvent can greatly impact the size and shape of the resulting crystals. Generally, slower evaporation results in minimal evaporation.

Conventional and Unconventional Crystallization Mechanisms

Edible oils and fats are essential nutrients in the diet as they provide essential fatty acids and energy. In addition, they have important functions in the production of processed foods. These ingredients are directly related to texture, stability, aeration, and nutritional features of the products. Some fats used in food are known by the term shortening, also known as plastic fats. Selection of the best shortening is based on formulations (product to which it is intended), functions, and process conditions. The production of shortening is related to the types of fats and oils used and the way in which they crystallize and form solid networks. This is because solid crystalline fat networks are responsible for providing the expected physical macroscopic functions of the various types of shortening systems, due to the way they melt and to the hardness, consistency, and complacency of the network.

Oils and fats are one of the three food macronutrients, represented by triacylglycerol (TAG) molecules, formed by an ester bond between glycerol and three fatty acids. TAG crystals are known to show polymorphism, which is the ability of a substance to have more than one structure or crystalline form. As the physical properties of fat crystals, such as melting point and rheology, are important for food texture, polymorphic control of fat crystals is crucial in the food industry, and it is thus necessary to understand the crystallization mechanisms of TAGs.

Crystallization consists of an arrangement of the system as a result of a driven force, characterized by the total or partial restriction of movement caused by physical or chemical bonds between TAG molecules. Due to the complexity of this system, TAGs can be compacted in different ways. This crystallization has implications in the industrial processing of foods, since these depend on fat crystals in products like margarines, chocolate, spreads, baked goods, etc.

Lipids

Lipids are defined as compounds soluble in organic solvents and insoluble in water. These compounds are represented by oils and fats. Chemically, vegetable oils and fats are mainly composed of TAGs and may contain small amounts of diacylglycerols, monoacylglycerols, free fatty acids, phospholipids, waxes, and unsaponifiable constituents.

Glycerol has three positions in its molecule (sn-1, sn-2, and sn-3) in which the fatty acids are esterified to form the TAG. The features of these fatty acids influence the physical properties of oils and fats: (i) in relation to unsaturation and saturation: saturated fatty acids have a higher melting point compared to unsaturated ones, since unsaturation impairs the packing of molecules; (ii) in relation to the geometric isomerism: it refers to the possible arrangements of the hydrogen atoms around the double bond, presenting the cis or trans form. Cis bonds reduce the melting point due to folding of the molecule, impairing packing. Conversely, fatty acids in a trans configuration have a melting point close to that of saturated fatty acids, due to torsion in the molecule, making it linear; and (iii) in relation to chain size: the longer the chain, the higher the melting point. Physically, oils are liquid at room temperature, while fats are solid, semi-solid, and plastic at room temperature. Therefore, the composition in TAGs determines the physical and functional properties of oils and fats, affecting the structure, stability, taste, and the sensorial and visual characteristics of foods.

Crystallization Mechanisms

Conventional Crystallization

Crystallization, melting, and recrystallization of lipids are first order transitions of interest in food systems that affect the shelf life of food and food ingredients. The crystallization properties of food components may be highly complex due to soluble-solvent interactions, mixture of miscible components, and polymorphic transitions. The production of solid fats, spreads, and chocolates is an excellent example of the use of first-order transition behaviors of food systems to manipulate properties in processing, structure formation, and application performance. Understanding the development of structures and the control of the crystallization of food components has evolved over the years, but there is still much to learn.

In lipids, the conventional crystallization process is fundamentally important to understand what happens while processing foods such as chocolates, spreads, bakery products, margarines, ice cream, etc. How can these crystals have such a great influence on the quality of a product? We can propose ice cream as an example. When consuming an ice cream, we want it to be creamy without crystals leaving a sandy sensation in the mouth. Another classic example is chocolate, as a properly crystallized cocoa butter of good quality, which provides a shiny and smooth chocolate that melts homogeneously.

Macroscopic properties of lipids in foods are particularly influenced by the microstructure of fats. Therefore, it is important to consider the effects of the microstructure for subsequent analysis of macroscopic properties. The formation and growth rate of the crystals and the polymorphic transformations are important to determine the function and application of oils and fats. Therefore, understanding crystallization is fundamental to obtain products with adequate functions, texture and sensory qualities and that are physically stable.

We will address the mechanisms that occur in vegetable fats during their crystallization. We will also understand that some techniques can induce or delay this crystallization in different ways, in order to achieve the industrially desired result.

Before discussing the conventional crystallization mechanism of vegetable fats, we need to define some concepts that are fundamental for a good understanding of the subject. We define conventional crystallization as an arrangement of the system characterized by total or partial restriction of movement, caused by the physical or chemical bonds between the molecules of TAGs. These bonds will lead to what we call crystals, which are molecules arranged in fixed patterns. Its high degree of molecular complexity allows the same set of TAGs to be compacted into several different and relatively stable structures.

We can say that the crystallization behavior involves four phases: nucleation (formation of a crystalline phase in the liquid phase through the organization of TAGs in a crystalline network, that is, molecules in the liquid state bind to create a stable nucleus), growth (caused by the inclusion of other TAGs in the existing crystalline network), recrystallization (structural and chemical change of the crystals during storage), and polymorphism (which controls the order of packing of TAG molecules in the crystals). Before forming the first crystals, the system, in its liquid state, needs to reach the supersaturated zone, in which there is a driving force for crystallization. From this point, crystallization starts.

Nucleation

It starts with the appearance of a driving force to overcome the energy barrier for crystallization. As a result, molecules in the liquid state somehow bind to create a stable nucleus. Nucleation can be classified as (a) primary nucleation (homogeneous or heterogeneous) and (b) secondary nucleation. Homogeneous primary nucleation occurs with the binding of isolated molecular species, which form dimers and trimers, and subsequently continues the accumulation process until a potential nucleus can be formed. That is, it is a molecular arrangement in a crystalline network, without external aid.

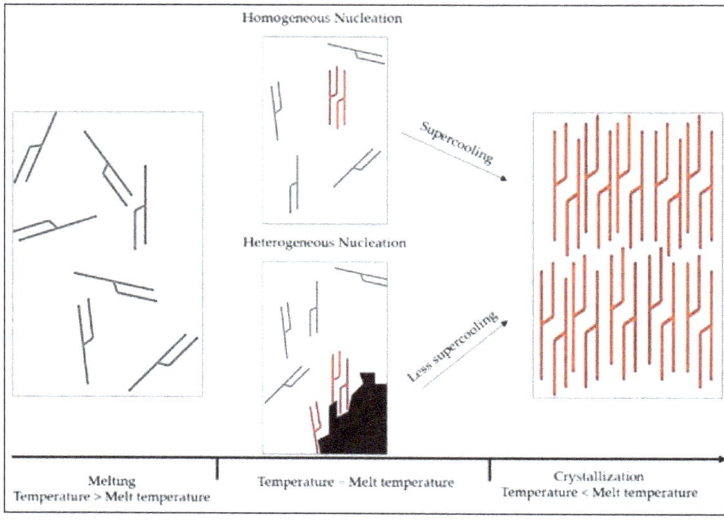

Nucleation mechanisms: crystal embryo formation in homogeneous nucleation and seeding or co-crystallization in heterogeneous nucleation.

Heterogeneous primary nucleation occurs when the local arrangement of molecules to form the nucleus arises from interactions between the solid particle and the supersaturated fluid, by supplying external energy. Secondary nucleation is the formation of a new nucleus in the presence of existing crystals, which may occur if crystalline microscopic elements are separated from an already formed crystalline surface, resulting in fracture of crystals in small stable nuclei. Once a crystal core is formed, it starts growing due to the incorporation of other molecules of the adjacent liquid layer, which is continuously filled by the supersaturated liquid surrounding the crystal.

Growth

According to the theory, crystalline growth occurs through two features: (1) surface of the growth process and (2) transport process in the face of the growing crystal. The mechanism is based on how the face of the crystal incorporates growth units of the adjacent solution. According to Hartman and Perdok, for a three-dimensional crystalline face, the capture of growth units can be classified according to how many binding interactions form between the adsorbed growth unit and the surface of the crystal. If three orientations of surface bonds are involved, the incorporation is then of the kinked face type; if two orientations of surface bonds are involved, the incorporation is of the stepped face type; and if only one surface bond is involved, the incorporation is of the flat face type.

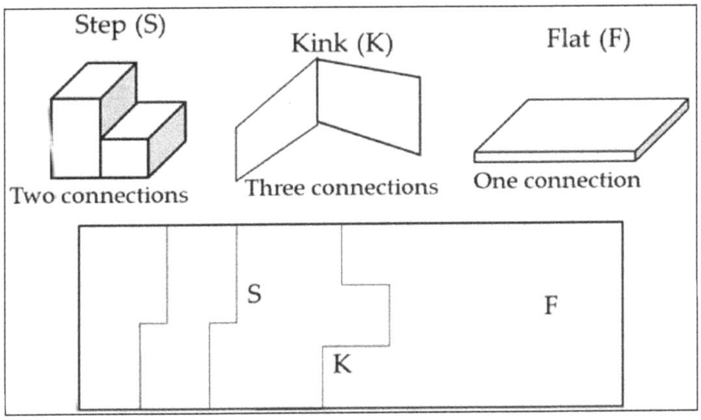

Classification of surface contacts according to growth theory.

Once a crystal nucleus is formed, it starts growing due to the incorporation of other molecules of the adjacent liquid layer that is continuously filled by the supersaturated liquid that surrounds the crystal. When the formed nuclei reach favorable dimensions, these elements become crystallites, and their growth depends on external (supersaturation, solvents, temperature, and impurities) and internal (structure, bonds, and defects) factors. Therefore, the crystal growth rate can vary by several orders of magnitude. Growth occurs through the binding of molecules to a crystalline surface. While the molecules are attached to the surface of a crystal, some molecules are also deactivated. There is a continuous movement of molecules at the surface of the crystal, and the result of these processes determines the growth rate, which is directly proportional to the subcooling and varies inversely to the viscosity system. Unless there is a kinetic constraint, growth continues until the system reaches equilibrium (driving force for crystallization of zero and maximum volume of the phase of the crystals).

Recrystallization

The final stage of crystallization in food occurs during storage, when crystals undergo a recrystallization step. This phenomenon can be defined as a change in the size, shape, orientation, and perfection of the crystals. Basically, in this stage, small crystals, which are more soluble, disappear and larger crystals grow. The concern with this step is related to the changes in the products during their storage, such as fat bloom in chocolates and coatings and oil exudation in fat-rich products.

Polymorphism

When discussing lipid crystallization, we must relate this phenomenon to one of the most important physical properties, which is polymorphism. Polymorphism can be defined in terms of the ability to present different unit cell structures due to several molecular packages. In lipids, there are three specific types of subcells (the smaller periodic structure that exists in the current cellular unit, which is defined as the transverse mode of packing of the aliphatic chains in the TAGs), the polymorphs: α, β', and β.

The α form is metastable, with hexagonal chain packing. The β′ form has intermediate stability and orthorhombic perpendicular packing. Crystals in the β′ form show greater function, since they are smoother and provide good aeration and creaminess properties. The β form has greater stability and triclinic parallel packing. The melting temperature increases with increasing stability (α → β′ → β) due to differences in molecular packing density. The crystal structures also differ in chain length, which illustrates the repetitive sequence of acyl chains in a unit cell along the axis of a long chain. The chain length structure ranges from double (DCL) to triple (TCL), fourth (QCL), or hexa (HCL). In addition, the conformation of the glycerol group also influences the phase behavior of fats. Two typical glycerol conformations, fork type and chair type, have been observed in TAG crystals. The fork-like conformation is favored by the presence of two identical fatty acids at positions 1 and 3 of the glycerol, forming a symmetrical TAG. On the other hand, the chair-type conformation is favored in asymmetric TAGs. Neighboring molecules of TAG crystals are stacked in layers, with the parallel hydrocarbon chains packed together and arranged more or less perpendicular to the parallel planes of the glycerol groups and to the methyl terminal group planes. The distances between these molecules characterize the short spacings and the long spacings, providing the polymorphic properties and melting point of each type of fat. The choice between the fork- or chair-like conformation is dictated by the need to accommodate fatty acids of different chain lengths and the distortions introduced by the double bonds.

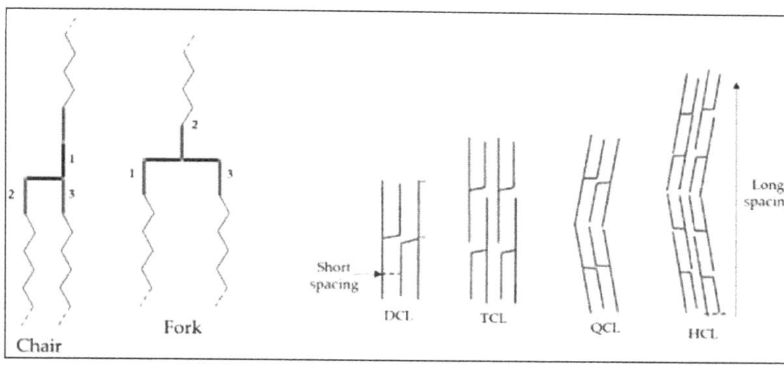

Conformation of glycerol groups in triacylglycerol crystals and chain length structures of triacylglycerol crystals.

Crystallization Modifiers

The crystallization process can be modified using techniques that change the kinetics of the crystals, such as the use of additives and ultrasound.

Additives

Lipophilic additives are used as they influence the volume properties (consistency and texture), promoting or inhibiting crystallization. These compounds provide a modeling effect, i.e., seeding nuclei. For example, a higher melting point additive with chemical and structural similarities to the lipid is used as a seeding nucleus for heterogeneous crystallization. Examples of additives that may promote or inhibit crystallization are minor lipids, such as diacylglycerols, monoacylglycerols, free fatty acids, phospholipids, and sterols. These lipids can interact with TAGs in the growth stage, causing a structural competitive effect or permanent incorporation into the crystalline structure, affecting the crystallization rate, polymorphic forms, and microstructure of the crystals, positively or negatively.

Another technique that is being increasingly used is seeding. It consists of the addition of a solid material with crystallization nucleating agent properties. The incorporation of crystallization seeds into liquid fats can promote two effects associated with the control of crystallization by the availability of numerous additional nuclei and/or surfaces for crystal growth. In addition, it may promote specific polymorphic forms.

The addition of emulsifiers changes the supersaturation characteristics of the solution and consequently its solubility, and thus, the incorporation of growth units on the surface will also change. At low supersaturations, growth occurs by the incorporation of the flat face type, and, in general, the free energy (ΔG) decreases according to the sequence kinked < stepped < flat. Thus, the mechanisms of co-crystallization and seeding occur, leading to the template effect of the emulsifiers.

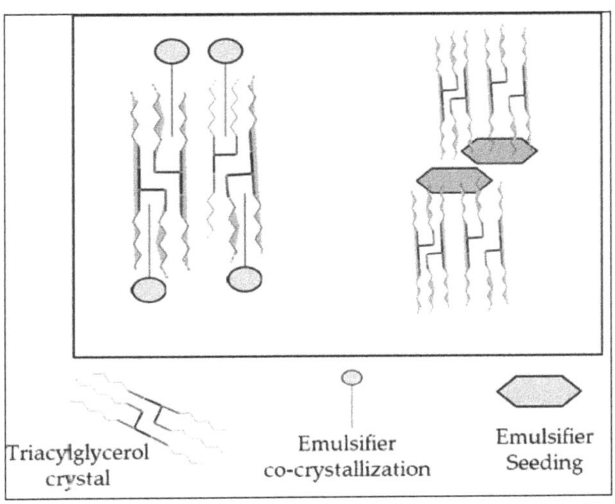

Template effect of emulsifiers: co-crystallization and seeding.

Ultrasound

This technique has been noted as an excellent alternative to promote crystallization, since the ultrasonic energy is sufficient to promote nucleation and growth by increasing the transfer of heat and mass. Through manipulation of nucleation and growth, ultrasounds can potentially provide improved control for the assembly of crystal structures in foods to control texture or improve separation. It is believed that these ultrasonic effects are superior to conventional agitation to control and stimulate the reaction rate, supersaturation level, nucleation, and crystal growth.

Care should be taken in relation to the intensity of ultrasound used, since the crystalline structure is highly dependent on the intensity of the ultrasound. For example, a palm oil where the ultrasound was applied at an intensity of 30 dB had little effect on kinetics but prevented the formation of large spherulitic crystals. When this intensity was increased to 35 dB, it produced smaller, more uniform crystals that tend to clump and fall to the bottom of the cell. A further increase in intensity to 40 dB, below the cavitational threshold, produced a uniform product of very small crystals, suggesting that there are numerous nucleation sites that produced crystals simultaneously.

The effects caused by the ultrasound include: (i) faster primary nucleation, which is quite uniform through the sonicated volume; (ii) relatively easy nucleation in materials that are otherwise generally difficult to nucleate; (iii) start of the secondary nucleation; and (iv) production of smaller, purer crystals of more uniform size. In addition, ultrasound can reduce clumping of crystals. The shock wave, which is caused by cavitation, decreases the contact between the crystals, preventing clumping, by controlling the population of local nuclei. The induction time is defined as the time elapsed between onset of supersaturation and appearance of crystals and decreases as supersaturation increases. This induction time is drastically reduced by the presence of sonication. This effect is more significant at low absolute supersaturations.

Unconventional Crystallization

Studies with organogels led to the development of many applications in the food area, such as chocolate filling, margarines, baked products, such as biscuits and cookies, pastries, spreads, sausages, ice cream, yogurt, unprocessed and processed cheese, mayonnaise, and sauces, among other products. Considering this, it is crucial to understand how unconventional crystallization occurs.

Many organogels are more desirable to consumers since they have a better nutritional profile than conventional fats. Therefore, organogel technology seems to be a potential solution for the reduction of saturated fatty acids and possibly the elimination of trans-fatty acids from processed foods.

This technology is feasible in comparison to the technologies currently used, since it does not cause any chemical changes in the structure of fatty acids and TAGs and maintains the nutritional features of the oil; in particular, it maintains unsaturated fatty acid contents and natural regiospecific distribution without increasing the AGS content.

Organogels are a class of materials that can hold large volumes of organic liquids in self-assembled networks and have many practical applications in the pharmaceutical, chemical, and food industries regarding lipid technology. Organogels are formed by structurants of low molecular weight and some organic solvents that are supported, through a three-dimensional network, forming the gel. Since organogels are biodegradable, they can be used in drugs as protein carriers and vaccines for example.

The formation of these structures through supramolecular assembly is of great interest, since they can form semisolid phases that are produced at low cost. Several researchers have focused their studies on the properties of structurants to form organogels.

In unconventional structuring, lipid systems composed of unsaturated TAGs, such as liquid or semi-solid vegetable oils, can be structured as gels, forming continuous networks of small molecules that assemble in liquid crystals, micelles, or fibrillar networks formed from aggregates of micelles, developing inverse bilayer structures in the form of rods.

Organogels can be structured forming a fibrous 3D network, where the solvent is trapped in the structuring matrix, avoiding the flow of solvent. The network is stabilized by weak interactions between the chains, such as hydrogen bonds, van der Waals forces, and π staking. Although it is known that organogels are formed through weak intermolecular interactions between the structurant molecules, which generate three-dimensional networks, there is still a lack of fundamental understanding of the type of interactions that are required.

The physically driven process of structuring organogels depends on many factors, such as structural adjustment of structurant molecules, solvent effects, concentration, temperature, use of ultrasounds, and shear. When these factors change, the structuring properties are affected due to the precise balance between the interaction of the structurant molecules, the interaction between the solvent and the structurant, and the applied external stimulus.

The structurants are insoluble in almost all solvents at room temperature, and the solvent-structurant and structurant-structurant interactions are the main force for the formation of organogels. However, they dissolve after heating and jellify after cooling; thus, the gelation process is an equilibrium between dissolution and precipitation in the solvents.

Organogel networks may range from the assembly of surfactants into solution by physical interactions (e.g., micellization, lyotropism, and crystallization), also known as molecular organogels, to the formation of flexible polymer networks (e.g., swelling), known as polymer organogels. The molecular organogels known today are at the interface between complex fluids and solids and, regardless of the nature of the structure, are composed of thermoreversible semisolid materials.

Structuring Modifiers

The group of molecular organogels can be subdivided according to the mode of self-organization of the structurant in liquid crystals, platelet crystals, elliptic networks, and inverted micelles.

Several solvent parameters have now been used to correlate with gelation ability, such as protic, dipolar aprotic, and apolar aprotic (low polarity), which were too broad to quantify solvent effects. More specifically, solvent parameters are divided into three categories: physical properties, solvatochromic properties, and thermodynamic properties, including dielectric constant, Reichardt ET-30 parameter, Kamlet-Taft parameters, Hildebrand solubility parameter, and Hansen solubility parameters.

The structures of the organogel molecules significantly influence the resulting gelation properties. The length of the alkali chain, the position of the chiral units, the substitution groups, and the peptide sequences, in addition to the arrangement or the number of the hydrogen bonds, may influence the properties of organogels.

The structures can be formed from amphiphilic compounds that can self-assemble in different microstructures, such as micellar and bilayer phases, which may accumulate in different lyotropic mesophases of larger scales (hexagonal, cubic, or lamellar liquid crystalline phases) at high concentrations. The formation of longer structures provides viscoelastic behavior or a gel-like solution. These viscoelastic materials, created through the assembly of structurants, are described as supramolecular organogels. The formation mechanism of these structures is based on the arrangement of monomeric units that bind through non-covalent bonds, such as van der Waals interactions, hydrogen bonds, electrostatic interactions, and π-π or τ-stacking.

Two different types of packing models were discovered in non-polar and polar solvents. In non-polar solvents, an interdigitated stacking model was preferred in the self-organized 3D micromorphology of the gels, due to inverse strong hydrogen bonds and weak π-π stacking interactions. Conversely, in polar solvents, the structurant is automatically assembled in a 3D nanostructure through a stacking model due to strong π-π stacking interactions and weak hydrogen bonds.

In general, there are a limited number of biocompatible components known to structure lipids through the assembly of molecules.

The extensive applications of organogels are of great importance to design and study new structurants and to characterize and determine the practical applications for organogels. It is known that some organogels with excellent mechanical and optical properties are more useful for practical applications.

Current knowledge on assembly behavior and subsequent gel formation and the effect of salts was applied to induce the formation of networks in hydrophobic environments using an unsaturated fatty acid, oleic acid, and sodium acetate, in sunflower oil. Oleic acid alone did not provide gelation in the oil, but the addition of sodium oleate induced gel formation. In mixtures containing oleic acid, reverse micellar structures are present. The participation of the sodium oleate in the original spherical micelles of oleic acid changes the geometry and assembly behavior of the structurant.

At higher concentrations of sodium oleate, lamellar crystal structures are formed. The assembly is theoretically driven largely by hydrophilic interactions between the hydrophilic head groups, as the strength of the interactions increased by the addition of small amounts of water. For water concentrations of more than 2%, the assembly was inhibited.

Vegetable waxes provide an ideal feature for organogels, which is more similar to the features provided by saturated and trans fats but show stability problems over prolonged periods.

There is currently a clear lack of knowledge regarding the identification of ideal proportions of ceramides, fatty acids, cholesterols, and phytosterols, in terms of oil structuring, crystal morphology, and consequent suitability for applications in foods. Although carbohydrate-based gelatins were extensively studied, there are very few studies in the field of organogels.

References

- Sunada, T. (2012), "Lecture on topological crystallography", Japan. J. Math., 7: 1–39, doi:10.1007/s11537-012-1144-4
- Definition-of-crystallize-605854: thoughtco.com, Retrieved 11 August, 2020
- Feng, Min; Yang, Aria; Zuo, Xu; Vittoria, Carmine; Harris, Vincent G. (2010). "Ab initio study on copper ferrite". Journal of Applied Physics. 107 (9): 09A521. Bibcode:2010JAP...107iA521F. doi:10.1063/1.3338905
- Conventional-and-unconventional-crystallization-mechanisms, crystal-growth: intechopen.com, Retrieved 08 June, 2020
- U. F. Kocks, C. N. Tomé, H. -R. Wenk and H. Mecking (2000). Texture and Anisotropy: Preferred Orientations in Polycrystals and their effects on Materials Properties. Cambridge University Press. ISBN 978-0-521-79420-6

Permissions

All chapters in this book are published with permission under the Creative Commons Attribution Share Alike License or equivalent. Every chapter published in this book has been scrutinized by our experts. Their significance has been extensively debated. The topics covered herein carry significant information for a comprehensive understanding. They may even be implemented as practical applications or may be referred to as a beginning point for further studies.

We would like to thank the editorial team for lending their expertise to make the book truly unique. They have played a crucial role in the development of this book. Without their invaluable contributions this book wouldn't have been possible. They have made vital efforts to compile up to date information on the varied aspects of this subject to make this book a valuable addition to the collection of many professionals and students.

This book was conceptualized with the vision of imparting up-to-date and integrated information in this field. To ensure the same, a matchless editorial board was set up. Every individual on the board went through rigorous rounds of assessment to prove their worth. After which they invested a large part of their time researching and compiling the most relevant data for our readers.

The editorial board has been involved in producing this book since its inception. They have spent rigorous hours researching and exploring the diverse topics which have resulted in the successful publishing of this book. They have passed on their knowledge of decades through this book. To expedite this challenging task, the publisher supported the team at every step. A small team of assistant editors was also appointed to further simplify the editing procedure and attain best results for the readers.

Apart from the editorial board, the designing team has also invested a significant amount of their time in understanding the subject and creating the most relevant covers. They scrutinized every image to scout for the most suitable representation of the subject and create an appropriate cover for the book.

The publishing team has been an ardent support to the editorial, designing and production team. Their endless efforts to recruit the best for this project, has resulted in the accomplishment of this book. They are a veteran in the field of academics and their pool of knowledge is as vast as their experience in printing. Their expertise and guidance has proved useful at every step. Their uncompromising quality standards have made this book an exceptional effort. Their encouragement from time to time has been an inspiration for everyone.

The publisher and the editorial board hope that this book will prove to be a valuable piece of knowledge for students, practitioners and scholars across the globe.

Index

www.ingramcontent.com/pod-product-compliance
Lightning Source LLC
LaVergne TN
LVHW080634210926
848485LV00003B/263